About
Body Mechanics for Manual

"Barbara Frye has taken a very complex subject matter and made it simple yet clear. A resource that all manual therapists should read and implement into practice, to enhance the quality of their experience in the treatment room."

—*Clint Chandler, L.M.P., C.N.M.T.*
Associate Faculty of Boulder College of Massage Therapy

"Physician: Heal Thyself! And that goes double for those of us who use our bodies to help others achieve their own healing potential. I applaud Barbara Frye's attention to the fundamental's of body mechanics and spatial awareness. I look forward to having a stack of these books in my office for immediate implementation by my manual therapist patients and colleagues!"

—*Lisanne Yuricich, D.C.,*
Lakeside Chiropractic

"At long last, a book that looks at "how" we use ourselves in working with others. The clear writing, repetition of key points, and copious illustrations make this book an excellent learning tool. By following the guidelines presented here, not only will you take better care of yourself, your clients will notice a dramatic increase in the quality of your interaction with them."

—*Mary Kay Hausladen,*
P.T., Guild Certified Feldenkrais Practitioner[CM]

"It is exciting to finally have a quality textbook available for massage and bodywork practitioners that addresses all aspects of body mechanics and movement as a preventative practice. It is long overdue for our profession. Bravo!"

—*Dawn M. Schmidt, L.M.P.*
Director of Education Brenneke School of Massage

"Barbara Frye has created a text that is both interesting and exploratory. I plan to explore my own body mechanics with the functional exercise portions of this book for years to come. *Body Mechanics for Manual Therapists* is an excellent didactic and experiential text that should be utilized by beginners and seasoned practitioners alike."

—*Marty Ryan, L.M.P.*
Co-owner of Seattle Advanced Bodywork Associates

BODY MECHANICS
for
MANUAL THERAPISTS

A Functional Approach to
Self-Care and Injury Prevention

BARBARA FRYE, L.M.P.

Licensed Massage Practitioner

*Guild Certified Feldenkrais Practitioner*CM

Illustrations by

ROBIN DORN, L.M.P.

Licensed Massage Practitioner

Published by FRYETÄG Publishing
P.O. Box 2688
Stanwood, WA 98292
Toll Free: (877) 474-1001

Edited by
Greg Bolton, B.S., L.M.P.
gdbolton@earthlink.net

Graphic design and production by
Clare Parfitt Design
http://www.clpdesign.com

Illustrated by Robin Dorn, L.M.P.
(206) 322-3241

Printed by Consolidated Press, Seattle, WA

ISBN 0-970-05210-3

Disclaimer

*The publisher is not responsible for any injury resulting from any material contained herein.
The purpose of this book is to provide information for manual therapists on the subject of
body mechanics. This book does not offer medical advice to the reader and is not intended as
a replacement for appropriate health care and treatment. For such advice, readers should
consult a licensed physician.*

Table of Contents

This book is lovingly dedicated to:

*The memory of my grandfather, Ernest J. North, for
teaching me the importance of faith, patience and humor.*

The memory of Glenna Noiles for nurturing my free spirit.

*And to my mom and dad, Enid and Glen, for always
believing in me. I love you both dearly.*

Preface

Body Mechanics for Manual Therapists is designed to help you develop a sound, effective way to use your body in your work. Regardless of the style of manual therapy you practice, attention to body mechanics is necessary to develop your self-care, promote the longevity of your practice, and help you deliver the best care to your clients.

As you study this book, you will acquire skills that will help you to prevent injuries and increase your overall well-being. To assist you, *Body Mechanics for Manual Therapists* has easy-to-follow lessons that give you a kinesthetic experience of important concepts.

If you are a student or just beginning your career, *Body Mechanics for Manual Therapists* will be a positive addition to your studies, preparing you to embark on your career with healthy and effective body mechanics. If you are a manual therapist in practice for several years, this book will give you some new ideas to help you continue your career in a positive way.

Early in my experience as a student of bodywork I became aware of how important healthy body mechanics were to my body and my practice. Now, as a Guild Certified Feldenkrais Practitioner[CM] and a licensed massage practitioner, I am dedicated to helping my students and clients find more comfortable and effective ways of moving. As an educator of bodywork, I have always focused on helping my students and fellow instructors implement healthy body mechanics.

Over the years I have become aware of how many manual therapists experience occupational discomfort and injury. This realization has been the driving force behind this book. My hope for you is that you will benefit from adopting the concepts presented here, and ultimately find pleasure and ease with your work!

Body Mechanics for Manual Therapists takes a very complex subject and makes it easy to understand and experience. So, have fun. Here's to your health!

Acknowledgements

I am thankful to God for the universal support given to me throughout this project.

This book would not have been possible without the help, support and generosity of many extraordinary people. I am deeply thankful to:

Robin Dorn for bringing her artistry, delight and magic to this book. Thank you Robin for your dedication and for helping to create the best flip book ever flipped!

Clare Parfitt for her graphic design and bringing to life the book of my dreams. And to Sydney for his ever present watchful eyes.

Greg Bolton for his editing and way with words, especially mine. Greg you were a gift from heaven.

Bob Seybold for the final proofread and creating an awesome index.

Diana Thompson and Angela Nacke for their assistance with the final editing.

Jackie Phillips for her generosity, support and loving brutality.

Andrew Biel for his generosity, encouragement and sharing of opinion.

Kate Bromley, Marissa Brooks, Paula Pelletier Butler, Leslie Grounds, Yvonne La Seur, Coleen Renee, Marty Ryan and Ann Wardell for meeting with me to work out the kinks. Thank you all for your feedback, suggestions and encouragement.

Kate Bromley, Kirk Butler, Paula Pelletier Butler, Robin Dorn, Denise Heinzmann, Yvonne La Seur, Lynnette Mathias, Melinda Maxwell, Angela Nacke, Carrie Nelson, Coleen Renee, Kay Rynerson, Annie Thoe and Roger Williams for their patience, creativity and support throughout the modeling process.

Clint Chandler, John Chester, Margaret Davidson, Nan Drake, Terry Graham, Elliot Greene, Laurel Harmon, Mary Kay Hausladen, Sean McDaniel, Ohbe Roe, Marybeth Saunders, Dawn Schmidt, Mary Short, Mitchel Storey, Katarina Stross, John Thoe and Lisanne Yuricich for their advice, support, feedback, encouragement and generosity.

Jerry Karzen and Jeff Haller for their guidance, wisdom and inspiration.

A special thank you to Jeff Haller for developing and teaching Use of Self. You fostered many of the concepts in this book.

My many students and clients over the years. Our experiences together made this book possible.

My family for their love and support.

Annie Thoe for her enthusiasm, support and vision. Professionally and personally you have deeply enriched my life.

Diana Thompson for her unwavering encouragement, support and listening ear. You have helped me to become a better writer.

Angela Nacke for her constant strength, patience and love. Thank you for teaching me die arschbacken zusammenkneifen—without it I would still be cleaning my house. And thank you for not going home with Tina Turner. Ich liebe dich.

And finally, Wasabi you are a constant source of joy, love and grace.

How to Use This Book

Body Mechanics for Manual Therapists consists of nine chapters. The first chapter, *Self-Care and Injury Prevention*, discusses body awareness and how developing awareness leads toward fostering health and longevity of practice.

Chapter 2, *The Basics,* discusses general aspects of body mechanics.

Chapter 3, *Tools of the Trade*, gives suggestions on how to best use your hands in order to keep them safe from injury.

The last six chapters are the functional chapters and include *Standing, Sitting, Lifting, Bending, Pushing and Pulling,* and *Applying Pressure.*

Each chapter begins with a brief introduction and a table of contents. Use the table of contents as a "chapter at a glance" reference. It may come in handy to find a concept or lesson quickly.

Chapters 3 through 9 begin with *The Habits of Everyday Life...* and *As a Manual Therapist...* These give you an opportunity to take a few minutes and think about how your everyday habits influence your body mechanics as a manual therapist. Spend some time with each of these pages and give each question some thought. By doing this, you will begin to develop body awareness. And as you will learn, developing awareness is the key to self-care and injury prevention.

Each functional chapter includes several key concepts that will lead you toward implementing effective and sound body mechanics. These concepts are often followed by a lesson called *partner practice* or *try this*. These lessons are designed to give you a kinesthetic experience of the concepts. After you have experienced the lessons, integrate what you have learned into your own specific type of manual therapy.

When going through the *partner practice* and *try this* lessons, keep a few things in mind:

◆ Take time and go through the lessons slowly. You can always go back and review them at a faster pace later.

◆ Always rest when you need to. There are plenty of rest breaks throughout the lessons, but if you need to rest before they come up, do so.

◆ Before beginning the *partner practice* lessons, let your partner know what you will be asking them to do. Give them the general idea of the lesson and make sure that they feel comfortable with what you will be doing. Whenever possible, practice with a variety of body types; clients come in all shapes and sizes.

◆ Be playful. If a movement feels strange or unfamiliar to you, don't take it too seriously. Just notice how it feels and move on, or come back to it later.

◆ Most importantly, allow yourself to discover what movements and body mechanics feel most effective for you. The lessons will guide you toward an effective use of your body, but you are the only person who can make the final choice of what feels right for you.

It is recommended that you read this book through at least once chapter by chapter; starting with chapter one, work each day as far as you can without losing your concentration and focus. Maybe you will work for 10 or 15 minutes each time you pick up this book. Some days you will feel like reviewing what you learned the previous day.

Don't push yourself. However you choose to use the book is fine. Take your time and go at your own pace.

The word "client" has been used throughout this book for simplicity sake. Don't be distracted by it if you use the words "patient" or "student." Use the word that fits your situation.

Key to Icons and Sidebars

Tips on self-care and injury prevention ──→

PREVENTION TIP

When lifting, keep your back in a neutral position. This will keep your back from becoming strained through the lifting process. You can encourage your back to stay in neutral by keeping your head up.

For Your Information— facts and curiosities ──→

CHECK THIS OUT

When holding a weight away from your body, your body perceives the weight to be at least ten times heavier then it actually is!

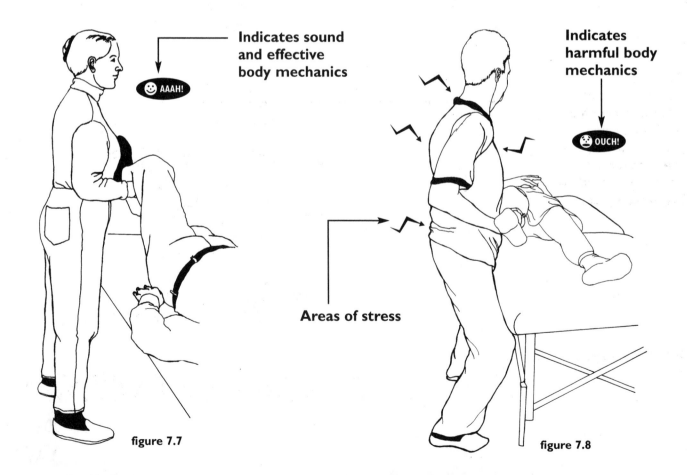

Indicates sound and effective body mechanics

😊 AAAH!

Indicates harmful body mechanics

😖 OUCH!

Areas of stress

figure 7.7

figure 7.8

Partner practice lesson

Lifting, holding and moving

Have your partner lie supine on your table. Stand beside your table, facing their lower leg. Make sure your entire body, including your feet, is facing their leg.

Tells you what position you and/or your partner will be in.

ACTION Slowly begin to lift their leg. Keep your back upright, bend from your hip joints and your knees. **(7.11)** As you lift their leg, straighten your legs. Lift and lower it a few times.

Indicates movement

■ Sense your feet pressing into the floor as you lift.

TIP By pressing your feet into the floor and straightening your legs, you are using the strength and power of your lower body to lift your partner's leg. **(7.12)** This allows your upper body to relax and comfortably facilitate the lifting without strain and effort.

Helpful information to consider, and when appropriate, integrate into the lesson.

ACTION As you lower their leg, bend your knees.

REST

figure 7.11

 Ask your partner for feedback.
✔ *How did using your legs affect the lifting of their leg?*

Partner feedback Be sure to take time and receive feedback from your partner. (This is an important part of your learning experience.)

figure 7.12

Self-discovery lesson

CHAPTER 1

Self-Care and Injury Prevention

INTRODUCTION

There is no time like the present to develop self-care, prevent injury and foster good health. There is, however, no fast and quick way to develop self-care. The process of learning how to maintain your health and longevity of practice is one that you must take on at your own pace and in your own personal way. The important point is that you make a commitment to yourself, and do it!

This chapter will set the foundation for the rest of the book, discussing the concepts on which each of the functional chapters are based. Here, you will begin to develop your body awareness by identifying your movement habits and sensing the difference between ease and effort. This will lead the way to continuing your development of self-care and injury prevention in each of the following functional chapters.

4

Body Awareness

"Awareness is the part of the consciousness which involves knowledge."[1]

—*Dr. Moshe Feldenkrais*

Before we can discuss specifics concerning your body mechanics, we must first discuss your body awareness. Body awareness is where it all starts. Without it, we blindly go through our work not realizing how our movements, responses, sensations and feelings affect our health. At this point, your level of awareness may be deep or you may be just starting to discover it. Whatever your level of awareness, becoming more aware of the truly extraordinary body in which you live is always a gift.

Self-care, which ultimately leads to injury prevention, begins with developing body awareness. Body awareness is a mindfulness of your body's movements, responses, sensations and feelings. As you develop this mindfulness or consciousness of your body, you become aware of subtle patterns, for example, the position of your shoulders when you read, or the shifting of your weight when you stand. Ultimately, developing body awareness requires you to become more observant of yourself, not only when you are performing manual therapy, but also during everyday life.

TRY THIS

Getting to know you

 Freeze yourself in the reading position in which you are presently.

■ You may be sitting, standing or lying down. Whatever your position is, hold it still for a few moments. You can continue to breathe, just hold your current position.

■ First notice your overall position.
  ✔ *Are you standing, sitting, lying down or in some other position?*
  ✔ *Is this a familiar position to you?*

■ Notice the position of your back.
  ✔ *Does this position feel familiar to you?*
  ✔ *Is your back comfortable?*
  ✔ *Does it feel sore?*

■ Notice the position of your legs and feet.
  ✔ *Does this position feel familiar to you?*
  ✔ *Are your legs crossed?*
  ✔ *Are your feet in contact with the ground?*
  ✔ *Are your legs and feet comfortable?*

■ Notice the position of your shoulders and arms?
  ✔ *Does this position feel familiar to you?*
  ✔ *Are your shoulders comfortable?*
  ✔ *Are you holding them up?*

■ Notice how you are holding this book.
  ✔ *Are your hands comfortable?*
  ✔ *Do they feel stiff?*

■ Notice the position of your head.
  ✔ *Are you holding it to the right?*
  ✔ *To the left?*
  ✔ *Does this position feel familiar to you?*
  ✔ *Does your neck feel comfortable?*
  ✔ *Does it feel tight?*

■ Notice how you are breathing.
  ✔ *Are you breathing mostly from your chest?*
  ✔ *From your abdomen?*
  ✔ *Does your current position allow you to breathe freely?*

■ Now stop holding your position.

■ Walk or move around a bit and take a couple of deep breaths.

Movement Habits

Once you become aware of your body, you can begin to become aware of your movement habits. Movement habits are patterns we repeat over and over again— often without being aware that we are actually using them. For example, the way you greet your clients, the way you answer the phone, how you use your hands to apply pressure, where you hold your head when standing and so on, are all elements of your movement habits.

In the previous awareness lesson, you had the opportunity to become aware of your reading position. You may have discovered certain positions that seemed familiar to you, like the position of your back, legs, shoulders or head. As we said, most all of our movements and postures are habitual and we reuse them over and over again. Once more freeze your position and notice how similar your position is now to the last time you froze. Even though you walked or moved around since the last lesson, chances are, you are still in a relatively similar position!

We need habits in order to live and because all of us are creatures of habit, we cannot help but transfer movement patterns and habits from one moment to the next and from one environment to another. As a manual therapist you transfer many of your lifelong and everyday movement patterns and habits into your working environment and guess where they show up—in your body mechanics. At the beginning of each functional chapter, you will have the opportunity to take a few minutes and consider your movement habits and patterns.

There are also movement patterns that you develop within your practice of manual therapy, habitual patterns that, for whatever reason, you repeat over and over. Becoming aware of these patterns will also increase your overall body awareness and guide you toward better self-care and injury prevention.

CHECK THIS OUT

As a baby, you developed postural patterns and habits that eventually evolved into the patterns and habits that you have today. Research has shown that at two months, a baby is forming postural patterns that include the movements of the eyes.

From the age of two months on, the baby forms postural patterns including facial expressions, sleeping positions, vocal patterns, and hand gestures. By the age of 2-3 years, the child is a "creature of habit."

TRY THIS

How your postural habits transfer from one environment to the next: Part 1

Stand as if you were going to talk to a friend whom you just ran into at the store.

■ While standing, answer the following questions for yourself:

✔ *Do you stand on both feet equally?*

✔ *Do you stand primarily on one foot?*

✔ *Which foot do you primarily stand on, your right or your left?*

✔ *Do you tend to stand still or do you shift your weight from one foot to the other?*

✔ *Do you keep both of your knees straight or slightly bent?*

✔ *Do you keep one knee straight and one knee bent?*

✔ *Do you bear more weight into one hip?*

✔ *Do you stand with one hip back or forward?*

✔ *Do you stand with your back vertical?*

✔ *Do you stand with your back slouched or slumped forward?*

✔ *Are your shoulders relaxed or are they held up, forward or back?*

✔ *Do you cross your arms in front of your chest or abdomen?*

✔ *Do you keep your hands in your pockets?*

✔ *Do you hold one hand on a hip?*

✔ *Do you rotate or tilt your head to one side?*

✔ *Do you look primarily in one direction?*

✔ *Do you look to the right or to the left?*

✔ *What happens to your breathing when you stand?*

✔ *Do you hold your breath?*

✔ *Do you take deeper breaths?*

REST

Walk around and shake yourself out.

How your postural habits transfer from one environment to the next: Part 2

Now stand beside your therapy table, as if you were going to work with a client.

■ While standing, answer the following questions for yourself:

✔ *Do you stand on both feet equally?*

✔ *Do you stand primarily on one foot?*

✔ *Which foot do you primarily stand on, your right or your left?*

✔ *Do you tend to stand still or do you shift your weight from one foot to the other?*

✔ *Do you keep both of your knees straight or slightly bent?*

✔ *Do you keep one knee straight and one knee bent?*

✔ *Do you bear more weight into one hip?*

✔ *Do you stand with one hip back or forward?*

✔ *Do you stand with your back vertical?*

✔ *Do you stand with your back slouched or slumped forward?*

✔ *Are your shoulders relaxed or are they held up, forward or back?*

✔ *Do you rotate or tilt your head to one side?*

✔ *Do you look primarily in one direction?*

✔ *Do you look to the right or to the left?*

✔ *What happens to your breathing when you stand?*

✔ *Do you hold your breath?*

✔ *Do you take deeper breaths?*

✔ *What standing habits did you find similar while standing in an everyday life situation and standing as a manual therapist?*

Ease vs. Effort: Your Skeleton

Once you become aware of your body and your habits of movement, you can begin to discover which movement habits serve you and which ones hinder and cause you discomfort, pain and injury. It is therefore important to learn why some habits of movement feel easy and why some require effort.

Gravity and how your musculoskeletal system responds to it are big reasons. Gravity is a force that we all must live with. Realizing that gravity can be used to your advantage is a crucial part of developing effective body mechanics and fostering self-care. When your body is working with gravity instead of against it, flexibility, endurance, balance and sensitivity are increased.

Your skeleton is designed to support your body to endure the force of gravity. It can best do this when it is used optimally—vertically aligned. When your skeleton is vertically aligned, all of your bones are "stacked" in such a way that they can endure up to around 2000 pounds of atmospheric pressure. In this case, your skeleton is being used as it was intended, for support, and your postural muscles can move your body with ease and comfort. However, when your skeleton is not used optimally, e.g., bending from your back instead of your hip joints, your postural muscles must work hard to take over for its lack of support. This causes a sense of effort in your body mechanics.

Gravity
by
Sir Isaac
Newton

The strength of bone

◖◗ Sit on a chair with your feet slightly in back of or in front of your knees.

■ Have your partner slowly sit on one of your knees. **(1.1)**

■ Notice how heavy they feel.
    ✔ *Can you feel the muscles of your leg working hard to support their weight?*

◖◗ Now sit with your ankle and heel under your knee.

■ Have your partner sit on your knee again. **(1.2)**

■ Notice if they feel less heavy then before—in fact you may not feel much of their weight at all.
    ✔ *Can you feel how the muscles of your leg do not need to work hard and that the bones of your leg are easily supporting your partner's weight?*

◖◗ Invite someone else to sit on the lap of your partner.

■ Notice again the strength you have in your leg, and that supporting this increased weight is relatively effortless.

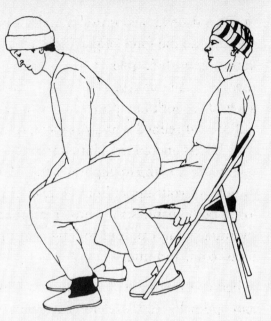

figure 1.1

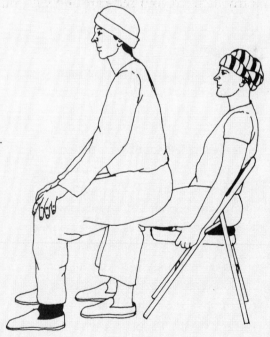

figure 1.2

Ease vs. Effort: Your Muscles

Your skeletal or postural muscles are voluntary muscles that keep you moving despite the force of gravity. The fact that they are voluntary means that you have control over their quality of movement. When they are used in a balanced manner, there is a sense of ease and comfort in your movements. The larger muscles are allowed to work in a powerful way and the smaller muscles are allowed to perform skillful and refined work. An incorrect use of your postural muscles can lead to areas of tightness and contraction, and often causes your skeleton to become misaligned due to the accompanying imbalance. For example, frequently lifting with the smaller muscles of your upper body instead of using the larger, stronger muscles of your lower body manifests a sense of effort and discomfort in your body mechanics. Eventually, this can lead to fatigue of the smaller muscles, abnormal holding patterns, pain, and injury.

When you are aware of your body and using your musculoskeletal system together and optimally, your skeleton can support you and your postural muscles can remain free to move you without excessive effort. Becoming aware of which movement habits in your body mechanics require effort and which ones feel easy and comfortable will greatly increase your self-care and will consequently prevent injury.

TRY THIS

Feeling ease vs. effort

figure 1.3

Put an object that weights about 10 to 15 pounds on a table. First stand next to the object and pick it up. (1.3)
✔ *Does it feel easy or difficult to pick up?*

Stand a few inches away from the object and pick it up. (1.4)
✔ *Does it feel more difficult to pick it up from this distance?*
✔ *Can you sense the amount of effort your muscles are using?*

Stand close to the object and pick it up.
✔ *Can you sense that you are using less effort and that it feels easier to pick up from this distance?*

TIP Sometimes it takes only a few inches to feel the difference between effort and ease in your body mechanics!

figure 1.4

Choice

In the last awareness lesson you felt the difference between ease and effort. If you had the opportunity to pick up the object again, which way would you choose—standing a few inches away or standing close? Hopefully you would choose the easier and more comfortable distance!

Developing body awareness—identifying your movement habits and sensing the difference between ease and effort—leads to choice. You can choose body mechanics that are uncomfortable and cause you discomfort or you can choose body mechanics that are easy, comfortable and effective. Choosing sound body mechanics is an ongoing process. There is no quick fix that will automatically give you pain-free and wonderful body mechanics. However, by developing body awareness, you will have something that is even better, an inner wisdom that will continually guide you in your self-care and promotion of health.

Notes:

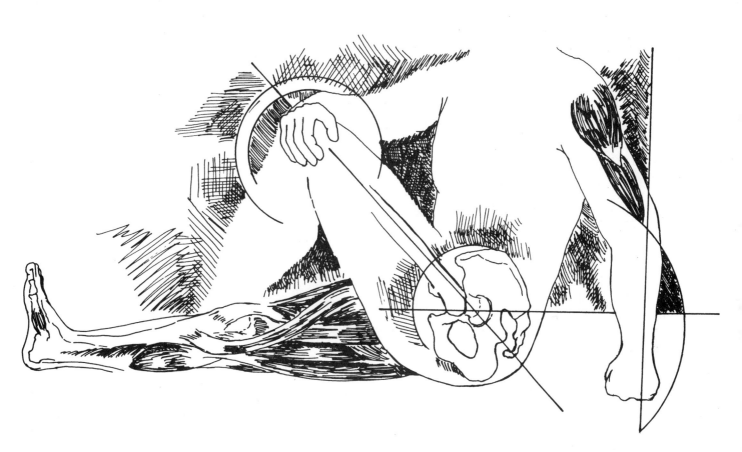

CHAPTER 2

The Basics

INTRODUCTION

This chapter includes general topics that are important to keep in mind before, during and after your sessions. We will discuss aspects of your environment, such as your table height and lighting; topics covering self-care, such as stretching, resting, breathing, pace and rhythm; and topics covering more personal concerns, such as your clothing and hair style. All of these are vital aspects that you will have the opportunity to explore and ultimately integrate into your practice.

Table Height

Your table should be set at a height that allows you to use your body weight rather than excessive muscular effort. If your table is too high, then your shoulders and upper body will be strained, and if your table is too low, then your low back will suffer.

There is no "correct" height for everyone. It is important for you to discover for yourself what is the best table height for you. Just because someone prefers a "lower" table, does not mean a "lower" table is right for you.

In general, if you stand to work, a good place to start is with your table at mid-thigh. This height takes into consideration your height and the depth of the client's body, and gives you an idea of whether or not you will be able to use your body weight. If you find that you need to adjust your table once you begin your treatment, and you do not have a hydraulic table, do not hesitate to stop and adjust it.

Ideally, as you become more comfortable and effective with your body mechanics, you will be able to adapt your body mechanics to any table height or situation.

If you primarily sit to work, your table should be at a height where your upper body does not need to strain in order to work with your client. Working with your table above your knees allows you to sit close and bend from your hip joints.

Other Working Surfaces

Though working at a table is the main situation used in this book, there are several other possibilities, for example, working with your client on the floor, standing while your client is seated, or getting on your table to work. (**2.1a, b, c, d**) Whatever your working surface is, take into consideration the points in each of the following functional chapters. Again, no matter what surface you choose to work on or what position your client is in, the principles remain the same. The more comfortable and effective your body mechanics, the less restricted you will feel.

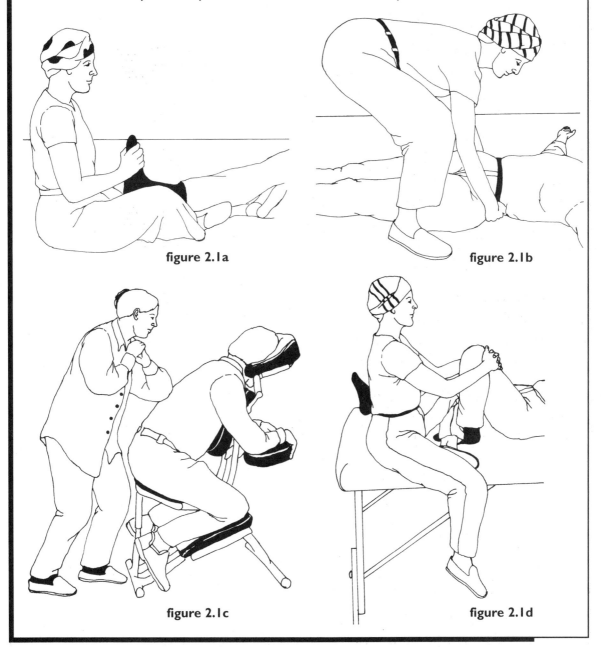

figure 2.1a

figure 2.1b

figure 2.1c

figure 2.1d

Chairs

A chair that gives you firm support is the best choice. There are many different kinds of chairs to choose from, but often they require your body to conform to their shape. This can cause your muscles to work very hard in order to keep a certain posture dictated by the shape of the chair. A chair that has a flat, firm and level surface gives your body the opportunity to support itself effectively and comfortably. Check out the chair you are currently using. If it has a sloping surface, then it could be forcing you to put too much weight on your sacrum. It could also be causing your pelvis to tilt back, which puts stress on your back. If your chair has a dipped, car-like bucket seat, it might be forcing you to mold your posture to its design. If it has a tilting backrest, again it may be causing you to put weight on your sacrum.

Chairs with back support are fine. Just make sure the back support is secure and does not automatically tilt back when you lean against it. As with the surface of your chair, you also want the back support to give you firm support. The height of the chair should allow your knees to be at the same height as your hips and your feet to rest flat on the floor. Have a chair available for yourself even if you do not sit to work—you never know when you might need to use it. This also gives your client the choice to sit before, during and after their treatment.

Lighting

Lighting should be bright enough for you to see what you are doing clearly. If the lighting is too low, your eyes will need to strain to see. This causes a chain of events starting from your head down to your feet. Your neck and head fall forward, your spine and pelvis follow and your legs and feet lose their stability. Low lighting can certainly create a relaxing and therapeutic atmosphere, but don't sacrifice your comfort and effectiveness. Give yourself enough light so you can maintain proper body mechanics.

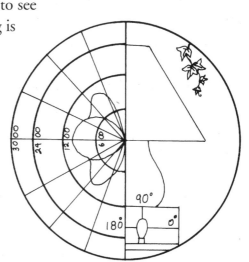

Floor Coverings

Rugs are wonderful for adding texture, but they can be problematic. If a rug or carpet is underneath or near your table, make sure it does not get in the way of your movement. It is easy to trip or become distracted by a rug which has a mind of its own. Keep it secured to the floor so it does not bunch up or cause you to interrupt your movement. Using a chair on a rug can also be distracting. Again, securing the rug so it stays in place while you are standing or sitting is ideal.

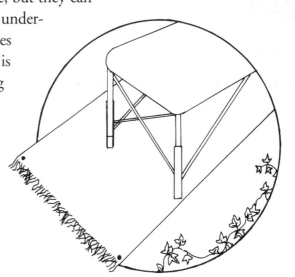

Space

Having enough space around your table is important. You need to be able to move around freely without worrying about bumping into something. If you are concerned about hitting a wall or piece of furniture, you cannot give your total attention to your work. If you are restricting your movements because of space, your body mechanics will be less than effective. On the other hand, not all room setup situations are ideal, and the key to working in a small space is to adapt to the space available. Clear the room of all unnecessary furniture and other items taking up space. Place your table so it has equal space around all sides. The most important point is to be relaxed and keep breathing!

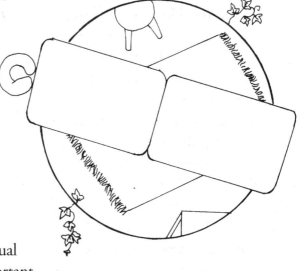

Clothing

Your clothing should not restrict or get in the way of your movement. If you need to stop frequently and rearrange your clothing, it is interfering with your body mechanics. There's nothing like having to stop in the middle of a manipulation to roll up a sleeve or tuck in a shirt. Also, if your clothes are too tight your effectiveness and comfort will be reduced. Wearing clothes that allow you to move freely is optimal.

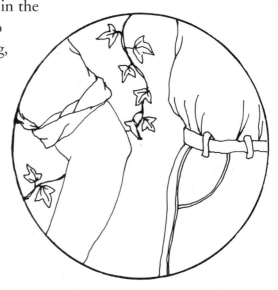

Shoes

Wear supportive and comfortable shoes. The comfort and support of your feet will reflect in your body mechanics. You may have experienced wearing uncomfortable shoes and can recall how they affected the rest of your body, not to mention your mood. Wearing non-slip shoes is also a good idea, especially on an uncarpeted floor. Your feet need to be connected and secure on the ground. Many therapists are comfortable working bare foot or in socks. This is fine as long as your feet feel good. If however, you find that your feet are constantly sore, wearing supportive shoes may be the answer. If you are the sock-wearing type, just make sure you are on a nonslip floor. You don't want to be sliding around your table like a deer on ice. There have been occasions where manual therapists have slipped and seriously injured themselves from wearing socks on a slippery floor.

Hair and Nails

Hair control can make all the difference in the world when it comes to body mechanics. Make sure your hair does not distract you while working. Keep your hair out of your face and prevent it from hanging down in front of you. Many manual therapists are not aware of how distracting their hair is until they find themselves holding their head to one side in order to keep it out of their face. Can you imagine how this would affect the neck and back? Believe it or not, the length of your nails can also affect your body mechanics. With short nails, there is no hesitancy to use the fingers or thumb. The hand can work in a soft and relaxed fashion. With longer nails, often the therapist is concerned with scratching or hurting the client. This results in tense and less effective hand usage.

Warming Up and Stretching

Manual therapy, no matter the type, is a very physical profession. It is therefore important to warm up with light exercise or stretching before starting your work. Increasing your circulation can help the responsiveness of your body during your treatments. If you start your work without warming up, your body is less responsive and more prone to injury.

Thinking of yourself as an athlete might help you to realize the importance of warming up. It would not make sense for a marathon runner to start a race without first warming up. Stretching or doing some light exercise increases her circulation and prepares her body for the stamina needed for running. The same is true for you. Warming up prepares your body for the stamina needed to do the work ahead of you.

Stretching between your treatments is also a good idea. Taking a few minutes between clients to stretch will help keep your body flexible and energized. Effective body mechanics will naturally reduce your muscular tension and increase your energy, but stretching will also help to keep your body feeling great!

The following stretches are a few of the many possibilities that increase flexibility and decrease tension.

When you begin, move to the point where you feel a mild and gentle stretch. Once there, be sure to breathe and relax as you hold the stretch for 30 to 45 seconds.

Do not bounce while holding a stretch and never stretch to the point of pain or discomfort.

wrist flexion

wrist extension

Arm, Abdominal, and Chest Stretches

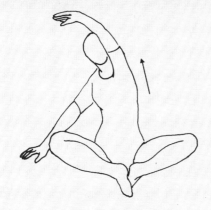

lateral trunk flexors

triceps

posterior shoulder

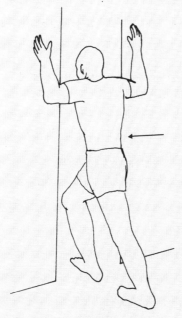

chest

chest

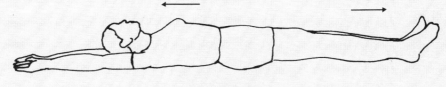

abdominals

Neck and Back Stretches

extension

flexion/extension

latissimus dorsi and shoulder

lateral side bends

back

flexion/rotation

back

Leg and Back Stretches

rotation

rotation

low back

low back

lower back and abductors

low back and hamstrings

Leg Stretches

quadriceps

groin

hamstrings

adductors

calf

Resting and Winding Down

During a long break or after your workday, take some time to unwind. Lie down on your table or better yet the floor and simply allow your body to rest. This gives your body and mind a chance to be still and relax.

There is a wonderful yoga pose called "the corpse." This pose allows your body to experience total relaxation. Here is a variation:

Lie on your back and place your arms beside your body, palms turned up. Keep your heels apart. Breathe slowly and deeply, and allow a sense of calm relaxation to come over your whole body. Focus on letting go of any tension.

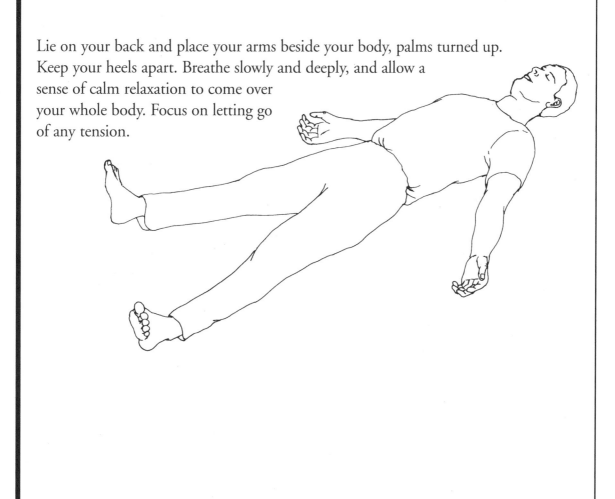

Breathing

Breathing is one of the most important aspects in developing effective body mechanics and preventing injuries. Having an awareness of your own breathing pattern will help you to realize when you tend to breathe and when you tend to hold your breath.

For example, do you hold your breath or breathe shallowly when you are talking to your clients? When you are mentally focused on a treatment area? Or when your body mechanics are uncomfortable? Sometimes, even initiating touch can cause the breath to become tense.

Consciously breathing deeply and slowly throughout your treatments benefits you in many ways: All of the systems of your body are nourished with oxygen, giving you more energy. Your internal organs are massaged by the movement of your diaphragm and between each inhale and exhale, your body is given the opportunity to rest, also giving you more energy. Ultimately, your body is more available for movement, which allows for unrestricted and effective body mechanics.

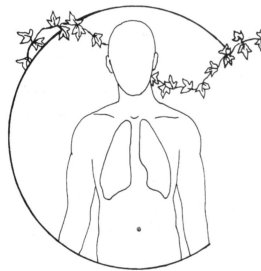

Breathing also gives your body a profound sense of relaxation while you work. For many reasons, remaining relaxed during treatments can sometimes be a challenge. However, when you consciously breathe slowly and deeply, your body and mind can tune into the pace of your breathing and ride along with it. The next time you find yourself becoming tense during a treatment or any other situation, take a few minutes to breathe slowly and deeply and allow yourself to calm down.

Pacing and Rhythm

As you know, movement is what makes your body mechanics alive and dynamic. Within your movement, you have your own pace and rhythm. It is important that you discover what your pace and rhythm are while working.

A therapist's unique pace and rhythm can often be what makes manual therapy enjoyable and interesting for them. Finding your pace and rhythm is discovering your own style and then riding it like a wave.

It is also important to acknowledge the pace and rhythm of your clients. For example, if you lift your client's leg quickly and your client is frightened by this, you might have interrupted their pace, putting them on guard. Finding a balance between your rhythm and pace and your client's rhythm and pace is ideal.

The next time you greet a client, notice how they move, talk and breathe. You will soon pick up on their unique pace and rhythm. This can help in relating to them as well as working with them.

The next time you work, notice what your pace and rhythm are. Do you have a steady pace and rhythm with no interruptions or do you have a kind of syncopated pace and rhythm with little pauses here and there. Increasing awareness of your unique pace and rhythm and acknowledging your client's will improve your working relationship.

PARTNER PRACTICE

Building awareness of the basics

Stand by your table, with your partner lying supine. Begin without touching your partner.

- First notice if your clothes are comfortable.
  ✔ *Are they distracting you?*

- Notice your hair.
  ✔ *Is it distracting you?*

- Notice if your table height is good for you.

- Notice if the space around you feels comfortable.
  ✔ *Do you have enough room?*
  ✔ *Is the lighting sufficient?*
  ✔ *Are there any rugs threatening to trip you?*

- Notice your breathing.
  ✔ *Can you sense how you are breathing?*
  ✔ *Are you breathing deeply or shallowly?*
  ✔ *Are you breathing slowly or quickly?*

- Sense from where you are breathing,
  ✔ *Are you breathing from your abdomen, your chest or from both places?*

ACTION Place one hand on your belly and one hand on your chest. As you breathe, fill your lungs with air and feel how your chest and abdomen move as you breathe.

ACTION Sowly begin to place your hands on your partner.
  ✔ *Notice if your breathing was interrupted by the movement of your hands.*

- With your hands on your partner, notice if you are feeling relaxed and at ease with yourself.

- If you are, become aware of why you are feeling relaxed.
  ✔ *What is it about yourself that makes you feel relaxed?*

- If you are not, become aware of why you are not feeling relaxed and at ease.
  ✔ *What is it about yourself that makes you feel uneasy?*
  ✔ *Can you help yourself to become more relaxed and at ease?*

ACTION Begin to move your hands on your partner—make movements that you would use in your manual therapy.

■ Again notice how your breathing is affected by the movement of your hands on your partner.
  ✔ *Can you continue to breathe deeply and slowly while you move your hands?*

■ Notice your pace.
  ✔ *Do you have a slow pace?*
  ✔ *Do you have a fast pace?*

■ Notice your rhythm.
  ✔ *Do you tend to move in a steady and uninterrupted rhythm?*
  ✔ *Do you tend to move a little and then stop?*
  ✔ *Do you move in a rhythm as if you were listening to music?*

■ Notice if your rhythm and pace match your breathing pattern.

ACTION As you continue to move your hands, become aware of your partner's rhythm and pace.
  ✔ *Can you sense how they are breathing?*
  ✔ *Can you see their chest or abdomen moving as they breathe?*
  ✔ *Are they breathing quickly or slowly?*

ACTION See if you can begin to match your pace and rhythm with their pace and rhythm.
  ✔ *How does this feel to you?*
  ✔ *Ask your partner how this feels to them.*

ACTION Now purposely change your rhythm and pace so that it is very different to that of your partner's.
  ✔ *How does this feel to you?*
  ✔ *Ask your partner how this feels to them.*

ACTION Once again, become aware of your breathing and allow yourself to relax.

TIP The basic concepts explored in this lesson can be integrated into each of your treatments. Become aware of yourself, your surroundings and the person you are working with. This will make a huge difference in your comfort and effectiveness and the comfort of your clients.

Notes:

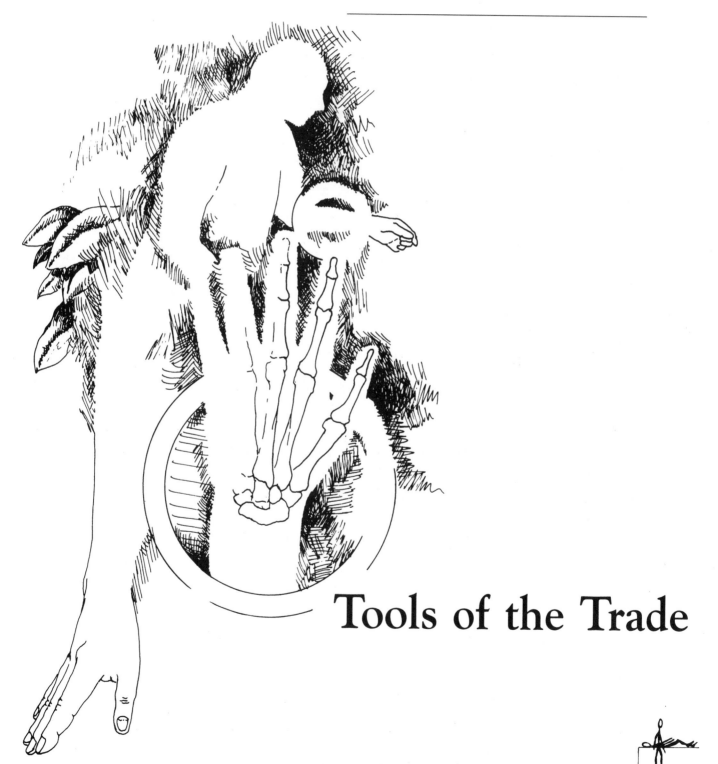

CHAPTER 3

Tools of the Trade

INTRODUCTION

There is no doubt that as a manual therapist, you use your entire body to carry out your work. However, your hands are your primary tools and it is therefore crucial to keep them healthy. Using them wisely will insure their health and the longevity of your practice.

The purpose of this chapter is to show you safe yet effective ways to use your hands.

You will have the opportunity to become more aware of how you currently use your hands and then learn some basic positions that will prevent the overuse patterns that lead to pain and injury. Keeping the positions and concepts you will learn in mind, you can integrate them into your own style and specific manual therapy. In this chapter you will also learn ways to effectively use your arms and will learn how to use your entire body to get the most from your hands.

No one but you can take care of your hands, and in doing so you will be insuring the health of your hands as well as your practice!

THE HABITS OF EVERYDAY LIFE...

How do you normally use your hands throughout the day?

◆ To shake hands with someone?

◆ To drive your car?

◆ To play an instrument?

◆ Do you grip tightly with your fingers?

◆ Are you aware of how you use your thumb?

◆ Your knuckles?

◆ In what position do you tend to hold your wrists?

◆ When do you use the palm of your hand?

◆ Your lower arm?

◆ Your elbow?

◆ Which is your dominant hand?

◆ Are you comfortable using your non-dominant hand?

AS A MANUAL THERAPIST...

How do you use your hands and arms?

◆ Are you aware of your wrists when you use your hands?

◆ Do you use your fingers to apply pressure?
◆ Your thumb?
◆ The palm of your hand?
◆ The heel of your hand?
◆ Your fist?
◆ Your knuckles?
◆ The ulnar side of your hand?
◆ Your elbow?
◆ Your forearm?

◆ How comfortable are you using your hands?
◆ Your elbows?

Your Wrist

Repetitive movements of the wrist are found in every form of manual therapy. Because your wrist is the connecting point between your hand and forearm, it is susceptible to overuse, pain and dysfunction. It is therefore vital for you to become aware of how you use your wrists and to keep them injury free. (Your wrists may be pain free now, but it is still important for you to take care of them to insure their future health!)

Ulnar and radial deviation are two movements that when held or used for long periods of time can cause great stress to your wrists. (**3.1a, b**)

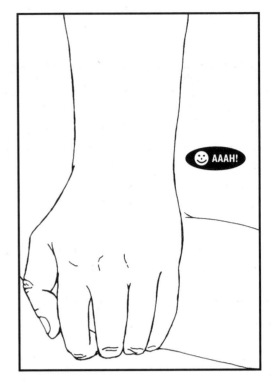

figure 3.2a

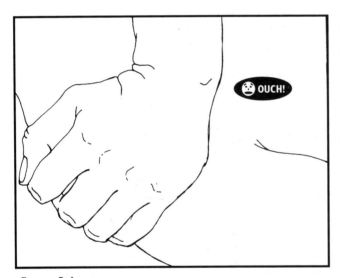

figure 3.1a

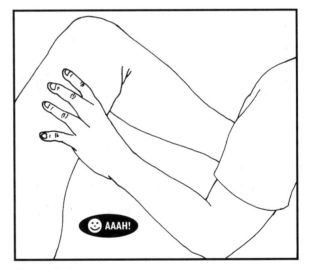

figure 3.2b

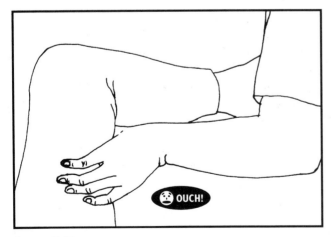

figure 3.1b

Keeping your hand and forearm in alignment with each other can decrease this stress. (**3.2a, b**) Often it is simply out of habit that the wrist is held in deviation. Bringing your awareness to the position of your wrists will make a huge difference.

Flexing and extending your wrist are natural movements when using the different aspects of your hand. However, holding your wrist in either position for long periods of time can cause stress, especially when applying pressure. Find ways to use your hand so that your wrist remains in a neutral position or at a slight angle. This will keep your wrist safe from pain and injury.

Extension of the wrist is most common and probably the most difficult to avoid. (**3.3**)

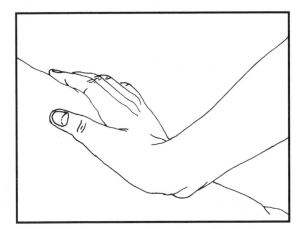

figure 3.3

Nonetheless, whenever possible find ways to decrease the angle between your hand and forearm, especially when applying pressure and pushing. (**3.4**)

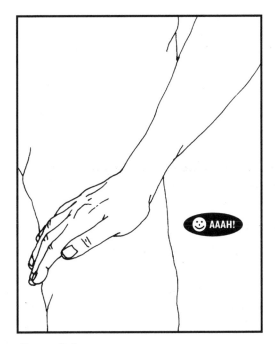

figure 3.4

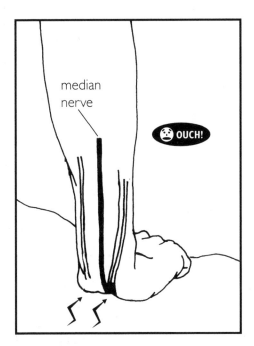

figure 3.5

Using your wrist in an extended position puts tremendous stress on the structures of your wrist. (**3.5**) You can avoid this stress by becoming aware of when you severely or with tension extend your wrist, and then finding alternative positions or ways to relax.

Flexion of the wrist is also common, especially when using the fist. (**3.6**) The fist is a great tool, but you should be careful not to hold your wrist in flexion for long periods of time. This position places stress on the wrist itself and on the extensor muscles of the forearm. When using your fist, find positions that allow your wrist to remain neutral. (**3.7**)

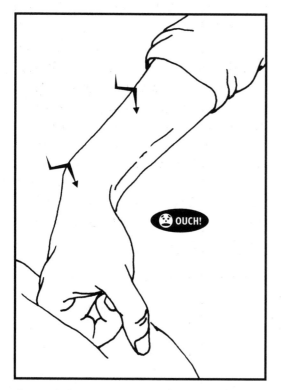

figure 3.6

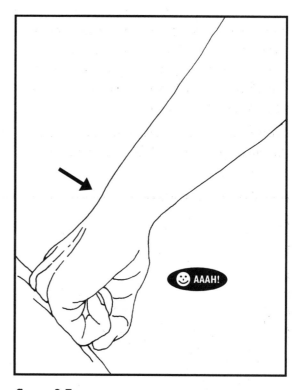

figure 3.7

CHECK THIS OUT

The bones of the fingers are called phalanges because they are arranged side by side, as were the Greek soldiers in the military formation known as the *phalanx*.

Your Fingers

The fingers give the hands dexterity and flexibility. They are made up of small slender bones that are divided by very small joints and held together by tiny muscles and ligaments. By design they allow us to grasp and grip. With them we can make tools, play instruments, groom ourselves—the list is endless. However, they are not designed to bear weight and, unfortunately, when asked to, are prone to injury.

As a manual therapist you use your fingers in a variety of ways. You receive information through them, you palpate with them and perform different kinds of manipulations with them. They are no doubt very valuable tools. Therefore, it is vital that you choose how and when to use them wisely.

Because of their delicate and flexible nature, the fingers are best used for palpation, grasping, gripping, and for non-weight bearing manipulations. While you may consider your fingers to be strong and capable of doing anything you ask them to do, if you use them in ways they were not designed to be used, they will become weak and susceptible to injury.

When using your fingers for palpation, use them gently and lightly. They can be amazing receptors for information if you use them sensitively and without force.

When using your fingers for grasping and gripping types of manipulations, keep your fingers as relaxed as possible. (**3.8**) If you grasp or grip in a stiff or strained manner, they will tire easily and quickly. Also keep in mind the flexor muscles in the forearm. (These are the muscles that move your fingers into flexion.) If you grasp and grip in a stiff or strained manner, the flexors become tight and strained as well. This is often what causes the forearm to feel tired and sore after use.

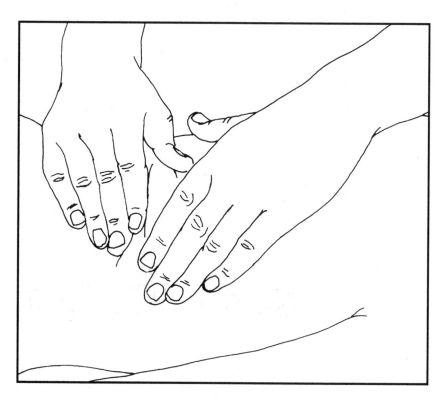

figure 3.8

44

The extensor muscles of your forearm, (the muscles that move your fingers into extension), are also affected by how you use your fingers. If the fingers are used in a forceful manner, the extensors become stretched and strained. Using your fingers in a relaxed yet meaningful way allows the extensor and flexor muscles of your forearm to work in a balanced manner.

Even though it is best to avoid using the fingers for weight-bearing manipulations, it is impractical for most manual therapists to actually do so. Therefore, the next best choice is to use them wisely. Do not apply pressure with your fingers for long periods of time and make sure the joints of each finger are aligned and supporting each other. (3.9) When the joints are not aligned, pressure and stress tend to build in the areas of misalignment. (3.10) It is also important that your wrist, elbow and shoulder are aligned when using your fingers. The more support you can offer them, the better.

Reinforce your fingers with your other hand to support them and guard against hyperextension. (3.11) It is very easy for the joints of the fingers to buckle under pressure. Avoid this by limiting their use for weight-bearing manipulations.

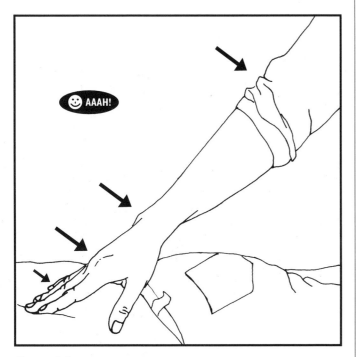

figure 3.9

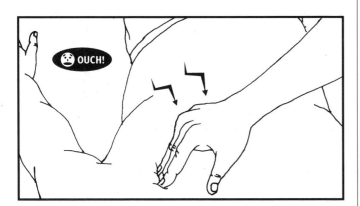

figure 3.10

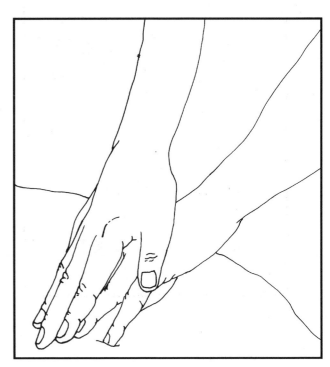

figure 3.11

Your Thumb

The thumb has a few advantages over the fingers. It has one less joint, which makes it more stable, and has several relatively thick muscles anchoring it to the palm. It also has the ability to oppose each finger, which allows it to participate in grasping and gripping activities.

Used by itself or with the fingers, the thumb plays an important role in palpation and manipulation. However, even though it has muscular support, weight-bearing manipulations put it at risk for injury. Many therapists avoid using the fingers for applying pressure and instead overuse the thumb. Though it may seem like the better choice because of its size and muscular base, if overused, the thumb like the fingers will become weak and ripe for injury.

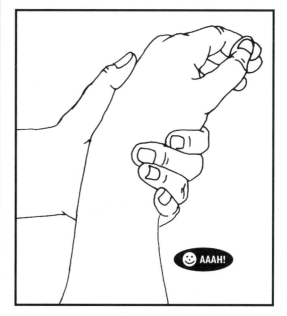

figure 3.13

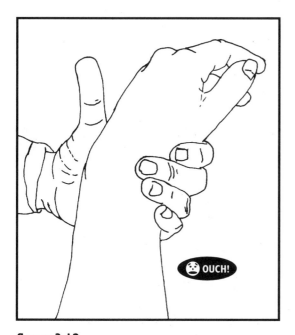

figure 3.12

It is common to see the thumb held in a static position when used with the fingers. (**3.12**) When using your thumb in combination with your fingers, allow your thumb to relax. (**3.13**) Though it employs several muscles, they are small and fatigue easily.

When using your thumb for applying pressure, reinforce it with your fingers and/or other hand when possible. (**3.14a, b**)

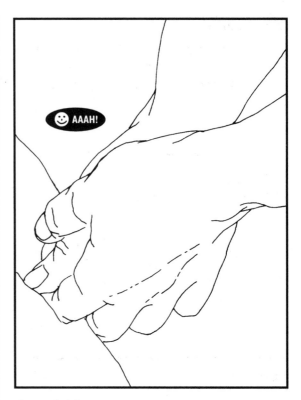

figure 3.14b

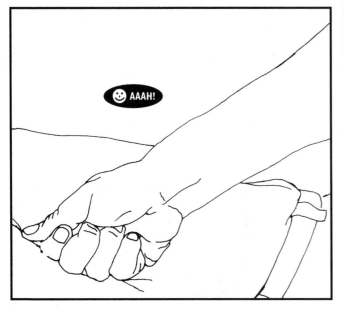

figure 3.14a

PREVENTION TIP

Use your thumb wisely. It has been shown that one pound of "pinch" between the thumb and index finger will produce six to nine pounds of pressure at the carpometacarpal joint of the thumb.

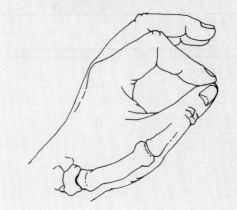

When using your thumb without reinforcement, do so for only short periods of time and make sure its joints are in alignment. (**3.15**) Lack of alignment can severely strain your thumb. (**3.16a, b**)

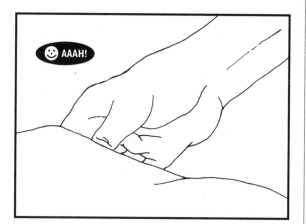

figure 3.15

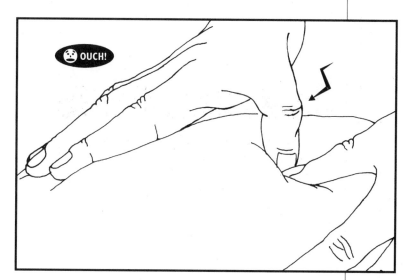

figure 3.16a

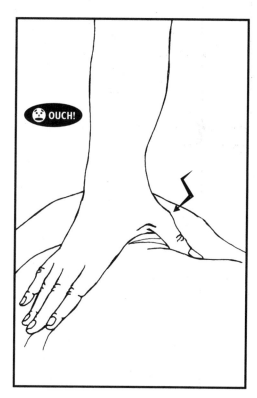

figure 3.16b

Your Palm

Your palm gives your hand shape and provides you with an excellent tool for light and deep manipulations. Using your palm allows you to touch a broader surface with the support of your fingers and thumb. The comfort and effectiveness of using your palm greatly depend on the angle at which you use your wrist and arm.

Keep your wrist and arm relatively aligned with your palm. (**3.17a, b**) This will decrease the chances of injury to your wrist when using your palm, especially when applying pressure.

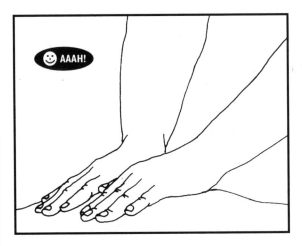

figure 3.17b

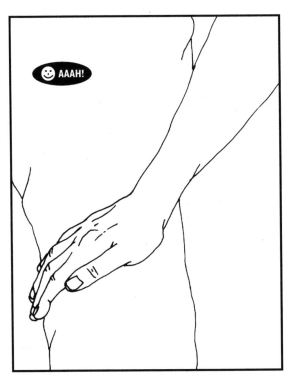

figure 3.17a

PREVENTION TIP

Use both of your hands equally to decrease injury. If you have the tendency to always use your right hand, for example, this overworks your right hand and puts it at risk for injury.

It may take a bit of time to become comfortable using your non-dominant hand but it will, in the long run, help to keep your hands injury free!

The Heel of Your Hand

The heel of your hand may seem like a great tool because it has a more bony prominence then the palm of the hand. However, the carpals, which make up these bony protrusions, provide passage for tendons and nerves.

Run your finger across the heel of your hand and feel for a little dip. This dip is the "carpal tunnel" where the median nerve shares space with several flexor tendons. Compressing into the heel of your hand can cause inflammation and injury to the median nerve and tendons. (**3.18**)

Therefore, incorporating the use of your palm along with your heel is the best choice. (**3.19**) This will help take some of the pressure off the carpal tunnel.

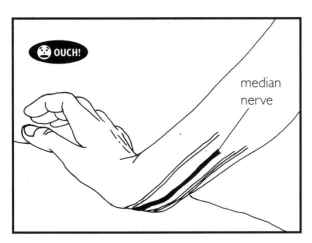

figure 3.18

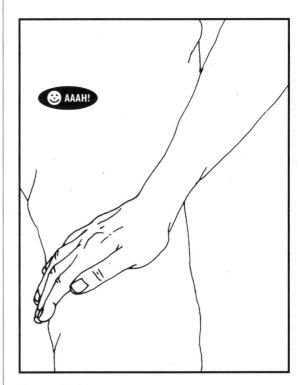

figure 3.19

Researchers have found fossils leading them to believe that human ancestors, living three million years ago, walked on their knuckles, just as chimpanzees and gorillas still do today.

Your Fist

The hand has the ability to curl the fingers into the palm and form a fist. A fist provides protection for the fingers and shapes the hand into a powerful tool.

The tendency when making a fist is to tightly hold the fingers inward, often creating a "white knuckle" effect. (**3.20**) The tighter the fingers are held, the harder the flexor and extensor muscles of the forearm must work—keep this in mind when using your fist. Don't find yourself over gripping your fingers. Let them fold in without creating undue tension in your wrist and forearm.

There are a variety of ways to use your fist. The top, flat part of your fist is a wonderful tool for weight-bearing manipulations, especially for applying pressure. (**3.21**) The ulnar and palmar sides are also good choices. No matter how you choose to use your fist, remember to keep your wrist and arm in alignment with it.

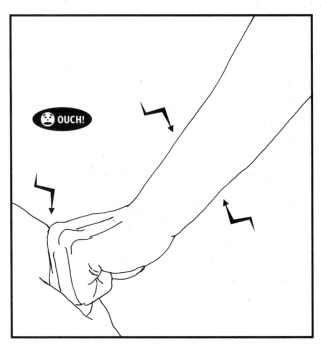

figure 3.20

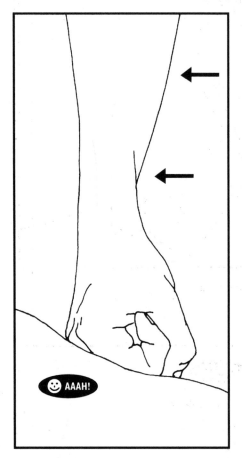

figure 3.21

Your Knuckles

Your knuckles are excellent tools and are a nice alternative to using the tips of your fingers. When used in proper alignment, they are more stable and stronger for weight-bearing manipulations.

Different aspects of your knuckles can be used, however, it is very important that you make sure your wrist stays in alignment with them. (**3.22a, b**) The tendency when using the knuckles is to flex the wrist while applying pressure, causing stress to the posterior and anterior side of the hand and wrist, as well as the wrist extensors at the lateral elbow. (**3.23**)

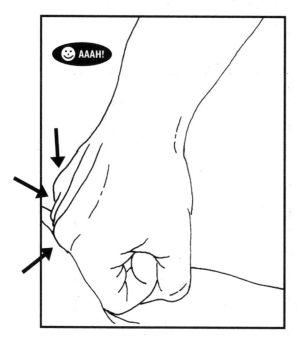

figure 3.22a

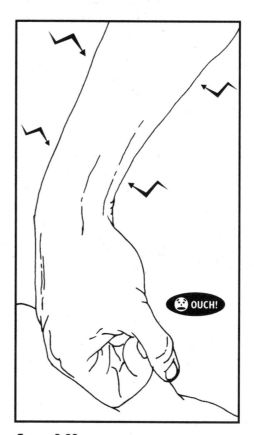

figure 3.23

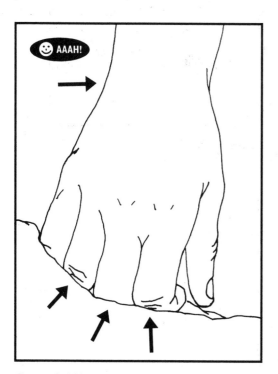

figure 3.22b

52

Your Ulnar Side

The ulnar side of your hand is a very effective tool. It can apply broader pressure and manipulate in the same way as your forearm, just on a smaller scale.

When using the ulnar side of your hand, keep your wrist in a neutral position. (3.24) Using your other hand to reinforce your wrist will help keep your wrist straight. (3.25)

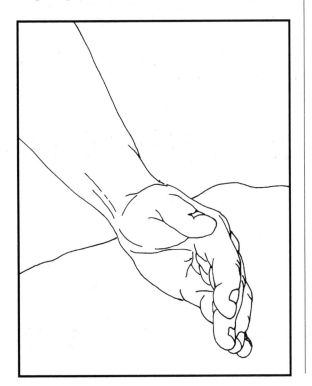

figure 3.24

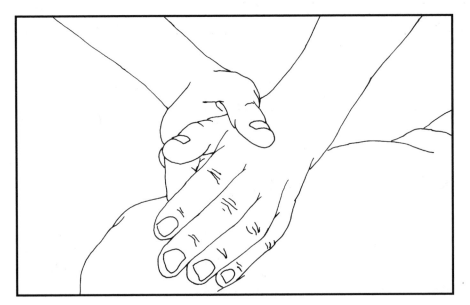

figure 3.25

Your Elbow

Your olecranon process or elbow is a terrific tool when used properly. It provides you with a strong and large bony process that can easily be used for weight-bearing manipulations. This is good news for your phalanges! Using your elbow can also decrease the overuse of your fingers and thumbs.

Your elbow is perfect for applying specific sustained and moving pressure. When it comes to working with thick, large and strong tissue, there is nothing better then the elbow.

Though it is a great tool, the elbow has little receptive ability. Therefore, the safest protocol is to first palpate the area of focus with your fingers or thumb before using your elbow. (**3.26**) Use one elbow at a time and keep your other hand free to guide it. (**3.27**) Think of your free hand as your elbow's eyes.

When using your elbow, keep your shoulder and elbow in alignment with each other, this will insure you are using strong skeletal alignment. Also, allow your hand to relax. Do not hold your hand in a tight fist. A relaxed hand will decrease fatigue in your hand, forearm and shoulder.

TRY THIS

Flex your arm and slide your finger across your elbow. Then slide your finger across the skin around your elbow. How much sensation does your elbow have compared to your skin?

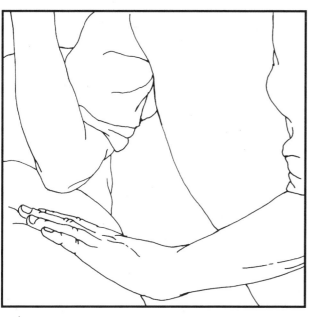

figure 3.26

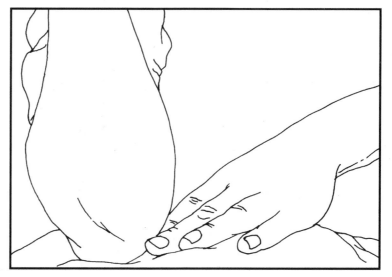

figure 3.27

Your Forearm

Your forearm is a great choice for applying a broader type of pressure. There are some forms of manual therapy which primarily use the forearm to manipulate the tissue. For example, Lomi Lomi, a Hawaiian tradition of bodywork, uses the forearms to apply a variation of strokes to the chest, stomach, back, buttocks, legs, feet, arms, and hands. This allows the therapist to cover a larger portion of the body while applying broad and deep pressure.

Using the ulnar or medial side of your forearm is the best choice of surface. (**3.28**) This helps to keep your forearm and wrist in a neutral position.

CHECK THIS OUT

A baby, at the age of three months, can lift her head and hold it up for one minute by supporting herself on her forearms.

AAAH!

figure 3.28

Using the anterior or posterior surface of the forearm requires you to hold your palm down in a pronated position or up in a supinated position. (**3.29a, b**) Either position requires effort and causes stress to the radioulnar joint and the muscles of the forearm and shoulder. (It is often in the radioulnar joint that many manual therapists experience discomfort and pain.)

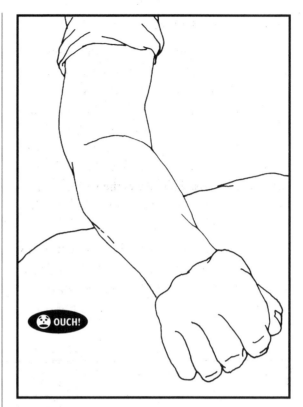

figure 3.29a

figure 3.29b

PREVENTION TIP

Keep your wrist and forearm in alignment when using hand held tools. Allow your fingers to grasp firmly, but do not overly grip the tool. Using other tools will not decrease stress to your own hand unless you use your own hand effectively to begin with!

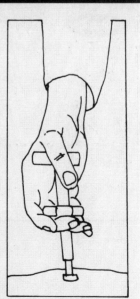

Using your forearm also allows your hand to rest. Therefore, it is important to keep your hand relaxed while using your forearm. It is common to hold the hand in a tight fist while using the forearm. (**3.30**) This does nothing but cause the flexor and extensor muscles in your forearm to fatigue quickly. Keep your fingers open and your hand relaxed to allow your arm to move more freely and with ease. (**3.31**)

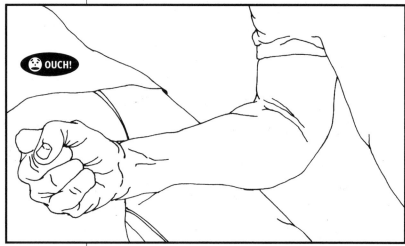

figure 3.30

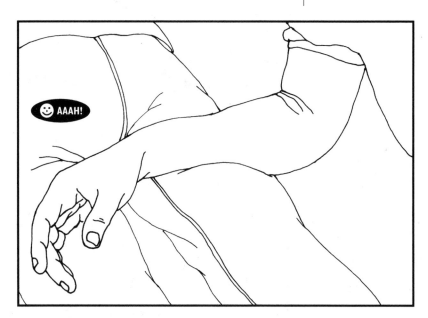

figure 3.31

PARTNER PRACTICE

Moving your body in symphony: Part I

Using your hands and arms wisely is crucial. Using them in coordination with the rest of your body is equally important. Integrate the movement of your entire body when using your hands to decrease stress and greatly increase the effectiveness of your body mechanics. When your hands and arms move together with your upper and lower body, your upper body is free to help facilitate the work without stress and your lower body can support you.

Stand by the side of your table.

ACTION Move the palms of your hands up your partner's leg. **(3.32)**

- Notice how you move yourself.
  ✔ *Do you move only your hands and arms?*
  ✔ *Do you move from your shoulders and back?*
  ✔ *Do you move your upper body and keep your lower body still?*
  ✔ *Do you place most of your weight onto your lead foot?*
  ✔ *Are you breathing normally?*

REST

ACTION Move your hands up your partner's leg once again, but this time only move your hands and arms, keeping the rest of your body relatively still. **(3.33)**

- Notice how this affects your body mechanics.
  ✔ *How do your shoulders and back respond?*
  ✔ *Does this feel like a familiar way in which you move your hands and arms?*
  ✔ *How does this affect your breathing?*

STOP *Take a break before continuing with Part 2 on the next page.*

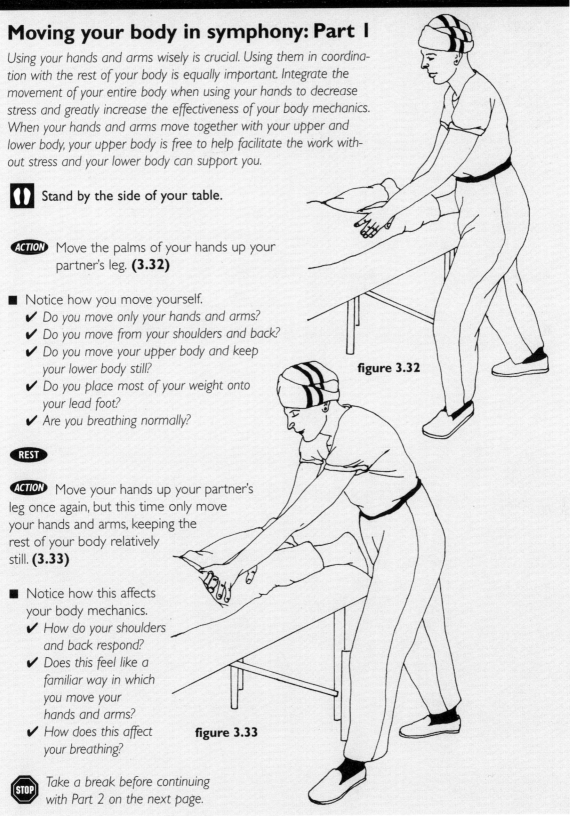

figure 3.32

figure 3.33

58

Moving your body in symphony: Part 2

Stand by the side of your table.

ACTION Move your hands up your partner's leg, this time moving your hands and arms first, then moving the rest of your body in the same direction. **(3.34)**

- Notice how this affects your body mechanics.
  - ✔ *How do your shoulders and back respond to this movement?*
  - ✔ *Does this feel like a familiar way in which you move your hands and arms?*

ACTION Now, move your body forward first and then move your hands and arms. **(3.35)**

- Notice how this affects your body mechanics.
  - ✔ *How does this movement affect your shoulders and back?*
  - ✔ *Does this feel like a familiar way in which you move your hands and arms?*

Finally, move your hands, arms

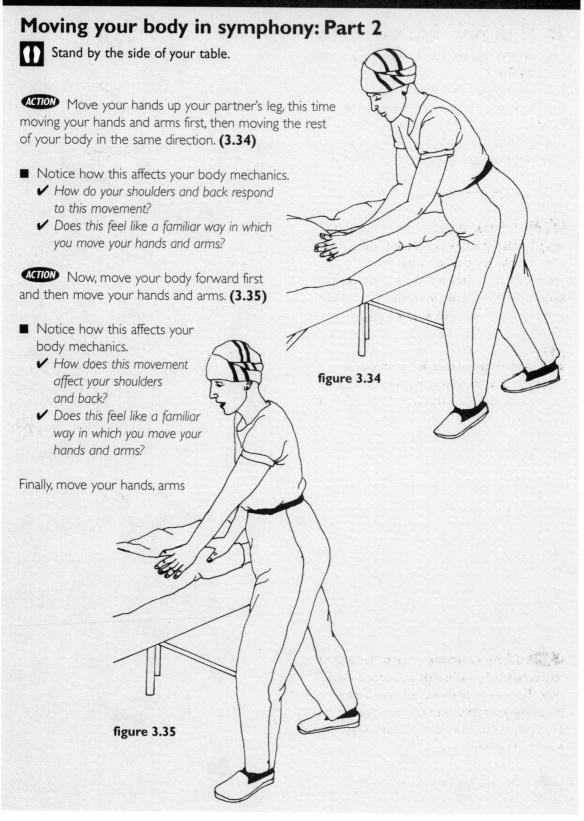

figure 3.34

figure 3.35

and the rest of your body together as you move your hands up your partner's leg. Allow your torso and lower body to help facilitate the movement of your hands and arms. **(3.36)**

✔ *How does this affect your shoulders and back?*

✔ *Can you feel how your muscular effort has decreased?*

✔ *Does this feel like a familiar way in which you move your hands and arms?*

✔ *Can you breathe more easily?*

TIP Transfer about 50 percent of your weight onto your lead foot and keep your rear foot in contact with the floor. If your weight cannot remain balanced between both feet, take a step up with your rear foot and then step up with your lead foot.

Ask your partner for feedback.

✔ *How did moving your hands and body in these different ways affect the quality of your touch?*

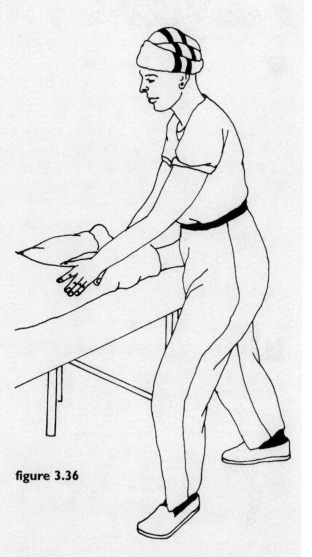

figure 3.36

TIP Use your entire body to facilitate the movement of your hands to reduce excessive effort in your shoulders and back. Involving your entire body also gives you the opportunity to use both of your legs and feet in an efficient manner.

TOOLS OF THE TRADE SUMMARY

1

Your wrist.
Find ways to use your hand so that your wrist remains in a neutral position or at a slight angle. This will help to keep your wrist safe from pain and injury.

2

Your fingers.
When using your fingers for grasping and gripping types of manipulations, keep your fingers as relaxed as possible. Do not apply pressure with your fingers for long periods of time and make sure the joints of each finger are aligned and supporting each other.

3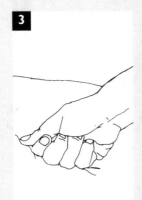

Your thumb.
When using your thumb in combination with your fingers, allow your thumb to relax. When using it for applying pressure, reinforce it with your fingers and/or other hand when possible.

4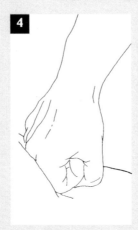

Your knuckles.
Your knuckles are excellent tools and are a nice alternative to using the tips of your fingers. When used in proper alignment, they are more stable and stronger for weight-bearing manipulations.

5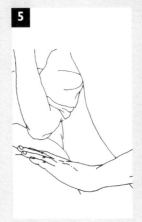

Your elbow.
Though it is a great tool, the elbow has little receptive ability. Therefore, the safest protocol is to first palpate the area of focus with your fingers or thumb before using your elbow.

6

Your forearm.
Use the ulnar or medial side of your forearm. This is the best choice of surface and helps to keep your forearm and wrist in a neutral position.

Notes:

Notes:

CHAPTER 4

Standing

INTRODUCTION

*S*tanding is a basic and foundational position from which all other body mechanics are performed. When you feel comfortable and supported in your standing, you will have a solid base from which to perform the more specific functions of your body mechanics.

In this chapter, you will learn how to support yourself using the tripods of your feet, with your legs in a strong angle of support, and vertically balance your upper body over your pelvis, legs and feet. You will also learn how to balance your head over your shoulders. Together these concepts will enable you to stand in an effective and comfortable manner.

THE HABITS OF EVERYDAY LIFE...

How do you normally stand throughout the day?

◆ When reading a magazine or newspaper?

◆ Talking to a friend?

◆ Waiting in a line?

◆ Do you stand with more weight on one foot than the other?

◆ Do you lean your weight into one hip?

◆ Do you stand with your feet wide or close together?

◆ Are you generally comfortable standing?

◆ Where do you experience discomfort when you stand for prolonged periods?

◆ Do you find yourself frequently having to change positions?

AS A MANUAL THERAPIST...

How do you stand?

◆ Do you stand with your feet in different positions?

◆ Do you primarily stand in one position?

◆ Do you place more of your weight on one foot?

◆ Do you distribute your weight between both feet?

◆ Is your upper body balanced over your legs and feet?

◆ Or do you brace yourself against your table?

◆ Do you tend to slouch forward?

◆ Is your head balanced over your shoulders?

◆ Do you lead with your head while standing?

68

Stand Still!

When you were young, did your parents or teacher ever tell you to "stand still"? When you are working, do you find yourself struggling to stand still? Here is some good news: it is next to impossible to stand still. Your body must continue to move or "sway" in varying degrees in order to keep itself in balance. This is how your body tells itself where it is in space. If it did not constantly move, you would literally fall over because your body would no longer have a reason to hold itself upright.

Department stores in the 1960's, for some reason, displayed fans blowing and holding up Ping-Pong balls. (4.1) As long as the fan was turned on, the Ping-Pong balls stayed in the air. But the minute the fan was turned off, the Ping-Pong balls dropped to the ground. The same is true for your body. As long as your body is in movement, you will remain standing. But if your body were to suddenly become still, you would fall to the ground.

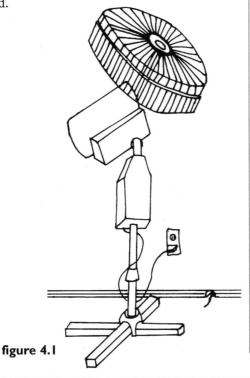

figure 4.1

CHECK THIS OUT

You may have the tendency to try and hold yourself still while standing. Hopefully you have not fallen over. But why haven't you fallen over? Your body will do everything possible to keep itself from falling. In fact 80% of your brain is occupied with keeping your body upright in space.

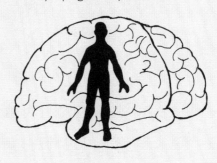

While you are standing and working, you are free to move and not expected to remain in a fixed or static position. The next time you feel yourself standing in a still or fixed position, allow yourself to move. Don't let yourself become so fixed in your standing that you begin to use too much of your energy just to remain standing. Remember, your body needs to move, so don't fight its natural movement.

PREVENTION TIP

You may find yourself standing with more of your weight on one hip and on one leg. This is common, especially when transitioning from one standing position to another.

There is nothing wrong with standing this way for short periods at a time. The trouble comes when this becomes the *only* way you stand. It puts a tremendous amount of strain on the weight bearing hip and leg.

Stand Straight!

Just as it is impossible to stand still, it is also impossible to stand straight. Yet, it is common to see people trying to stand "straight", with their shoulders back, chins up, and knees locked. (**4.2a, b**) Trying to stand "straight" takes your skeleton out of its natural alignment and requires your muscles to work very hard to hold a "straight" posture.

The next time you feel yourself trying to stand "straight", relax and allow the natural shape of your skeleton to support you. (**4.3a, b**)

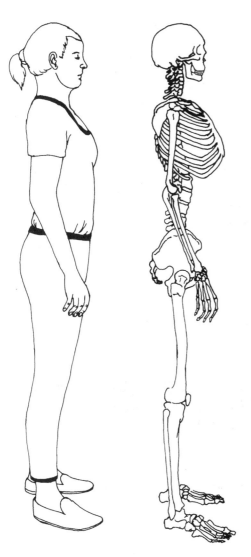

figure 4.2a   **figure 4.2b**

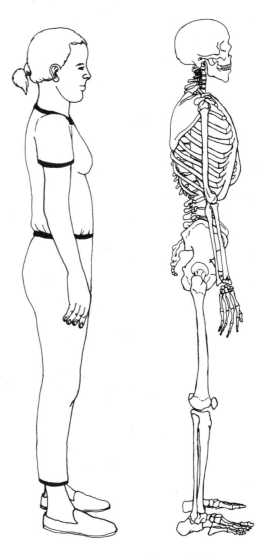

figure 4.3a   **figure 4.3b**

If you look at the human skeleton, you will find 206 bones, with many shapes; long, short, flat and irregular. But you will not find one straight bone.

Your skeleton has several curves and shapes that, with the help of your muscles, gives you dynamic movement, balance, shape and form.

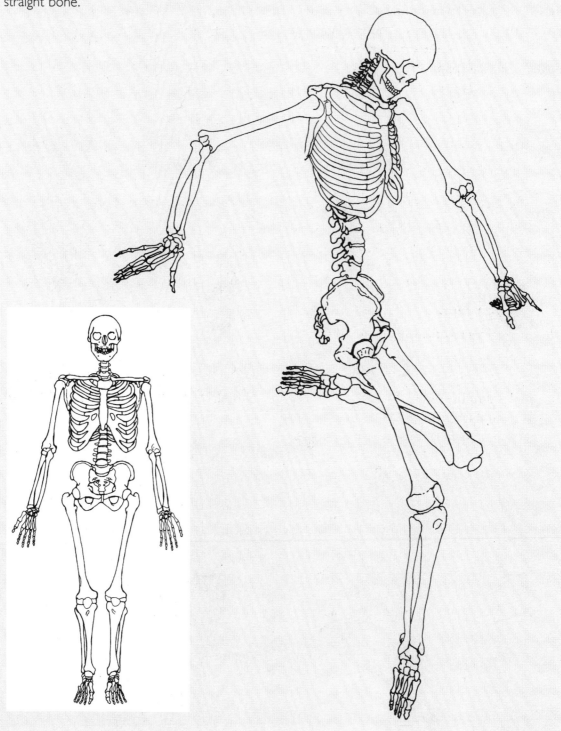

Your Bare Feet

Because your feet bear the entire weight of your body, they are constantly under an extraordinary amount of pressure. In fact, during a typical day, your feet endure a cumulative force of several hundred tons. It is therefore not surprising to experience fatigue in your legs and feet. When the structure of your foot is not used for support, the muscles of your lower body must work hard to support you. (4.4) This not only stresses the muscles, but also puts tremendous stress on the joints of your knees and ankles.

The soles of your feet have three points which support them when in contact with the ground and help to distribute your body's weight evenly. Together they form a triangle, or tripod, on which your foot finds balance when standing. (4.5)

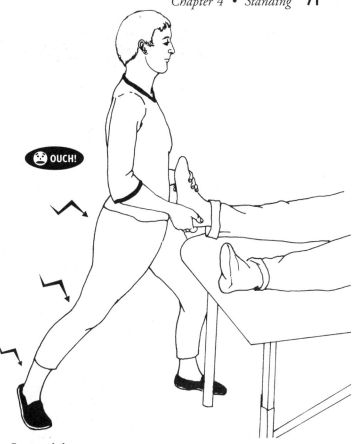

OUCH!

figure 4.4

These points are:

◆ the calcaneous tuberosity (your heel)

◆ the head of the first metatarsal (the base of your big toe)

◆ the head of the fifth metatarsal (the base of your little toe)

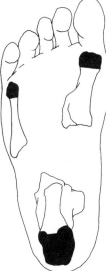

figure 4.5

CHECK THIS OUT

Many animals have tripod shaped feet which increases their balance and support when standing.

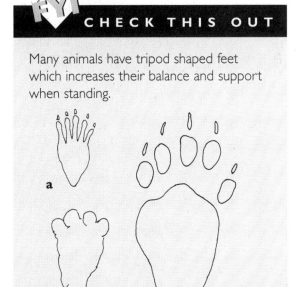

a

b   c

Animal footprints:
a. raccoon   **b.** snowshoe hare   **c.** grizzly

72

Your foot is a biological masterpiece. Contained within its relatively small size are 26 bones, 33 joints and a network of more than 100 tendons, muscles and ligaments, to say nothing of blood vessels and nerves...*and* your two feet together contain a quarter of all the bones in your body!

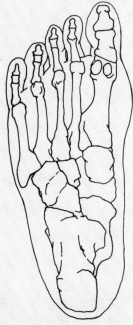

"A masterpiece of engineering and a work of art"

—Leonardo da Vinci

In order to stand with your weight evenly distributed on your feet, you must use your legs in a supportive and stable way. Whether you stand with your feet parallel to each other or with one foot forward, your legs must transmit the weight of your body down to your feet equally. In this situation your legs form a triangle which, according to the principles of engineering, is the strongest structural shape.

When standing with your body's weight evenly placed through your foot's natural tripod and with your legs in a strong angle of support, your entire skeleton has a solid base on which to stand. (4.7) The muscular effort of your lower body is decreased as is the stress on your knees and ankles.

This also greatly increases the flexibility of your feet and helps your feet to feel less tired after a day of standing.

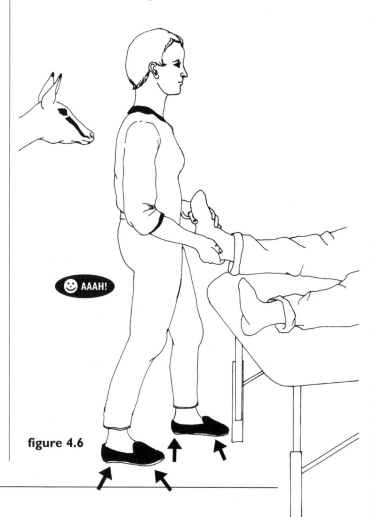

figure 4.6

TRY THIS

Standing on your tripods: Part 1

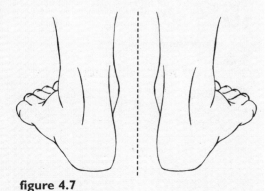

Take off your shoes and stand with your feet under your pelvis.

ACTION Distribute your weight equally between both feet. Look down between your feet and notice the distance each foot has from your center line.

TIP Your center line is a line that, if drawn from your head down between your feet, would divide your body into equal halves.

ACTION Place your feet at equal distances from your center line. This will create a strong angle of support from which your legs can transmit your weight equally to your feet. **(4.7)**

figure 4.7

■ Feel which parts of your feet you are standing on.
✔ *Are you standing with your weight equally placed on your feet?*
✔ *Are you standing more on your heels?*
✔ *Your forefoot?*

ACTION Lean your body back so that you stand primarily on your heels. **(4.8)**
✔ *Is this a position in which you normally stand?*

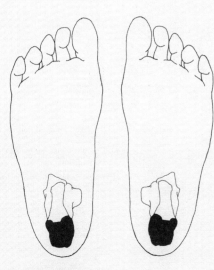

■ Notice how standing on your heels affects your lower body.
✔ *Do you feel the muscles in your legs working to hold your balance?*
✔ *Can you feel something happening in your knees?*
✔ *In your ankles?*

figure 4.8

■ Notice how this position affects your upper body.
✔ *Do you feel the muscles in your back working to hold your balance?*
✔ *What about your head and neck?*
✔ *How does this affect your breathing?*

STOP *Take a break before continuing with Part 2 on the next page.*

Standing on your tripods: Part 2

 Stand again and place your feet at equal distances from your center line. Distribute your weight equally between both feet.

ACTION Slowly lean your body forward and stand primarily on the balls of your feet. **(4.9)**

✔ *Is this a position in which you normally stand?*

■ Notice how standing on the balls of your feet affects your lower body.
  ✔ *How do the muscles in your legs respond?*
  ✔ *Can you feel your calves working hard to keep you from falling forward?*
  ✔ *How do your knees and ankles feel?*

■ Notice how this position affects your upper body.
  ✔ *How do the muscles in your back respond?*
  ✔ *What about your neck and head?*
  ✔ *Can you breathe comfortably?*

REST

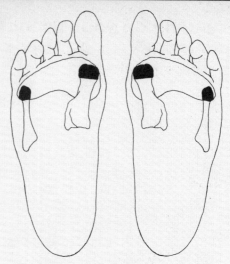

figure 4.9

Stand again and place your feet at equal distances from your center line. Distribute your weight equally between both feet.

ACTION Let your body weight travel down the strong angle formed by your legs and onto the supportive tripods of your feet. **(4.10)** Your weight will now be supported by your heels and the balls of your feet.

■ Notice how standing on the tripods of your feet affects your lower body.

figure 4.10

  ✔ *Has the muscular effort in your legs decreased?*
  ✔ *Has the stability increased in your knees and ankles?*

■ Notice how your upper body is affected.
  ✔ *Is your back more comfortable?*
  ✔ *What about your neck and head?*
  ✔ *Can you breathe more freely?*

Take a few minutes and continue to stand on your tripods. Sense your overall balance and support.

Goofy Foot

The tendency when standing with one foot in front of the other is to form a "T" with the feet, causing one side of the body to turn in the direction of the rear foot. (**4.11**)

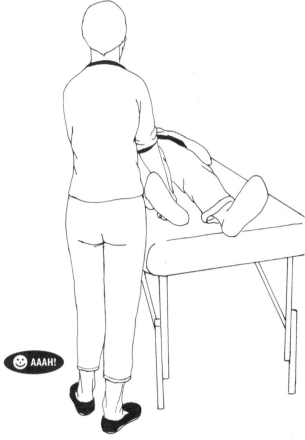

figure 4.12

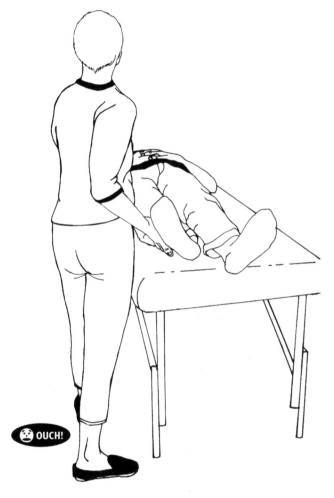

figure 4.11

Though a "T" stance may seem like a stable position, it creates a rotation or twist in your body. While one side of your body is directed forward, the other side of your body is turned to the right or left, away from your area and direction of focus. Standing with your feet in this kind of stance can cause stress to the side of your body that is rotated away. (The low back is usually where most of the stress is felt.) The knee and ankle of the rear foot are also at risk for injury because of their rotated position.

Instead, find a stable stance where both of your feet are directed, as much as possible, toward your direction of focus. This will decrease the risk of strain to your back, knees and ankles. Both sides of your body will be able to move without conflict, increasing your mobility and comfort. (4.12)

Remaining Vertical

When standing, it is important to keep your upper body vertical and balanced over your pelvis, legs and feet. When vertical, your center of gravity remains over your base of support. Keeping your center of gravity over your feet lets your upper body move without restriction, allowing your shoulders, arms and hands to move freely. (4.13) However, when your upper body moves out of vertical alignment, for example when you slouch or slump forward, your center of gravity falls away from your base of support. This requires the joints and muscles of your upper and lower body to work hard in order to keep you in balance. (4.14)

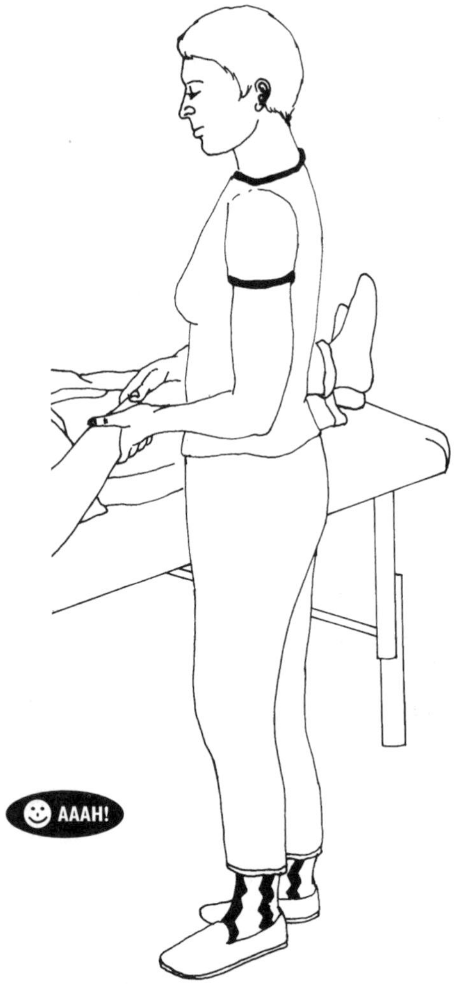

figure 4.13

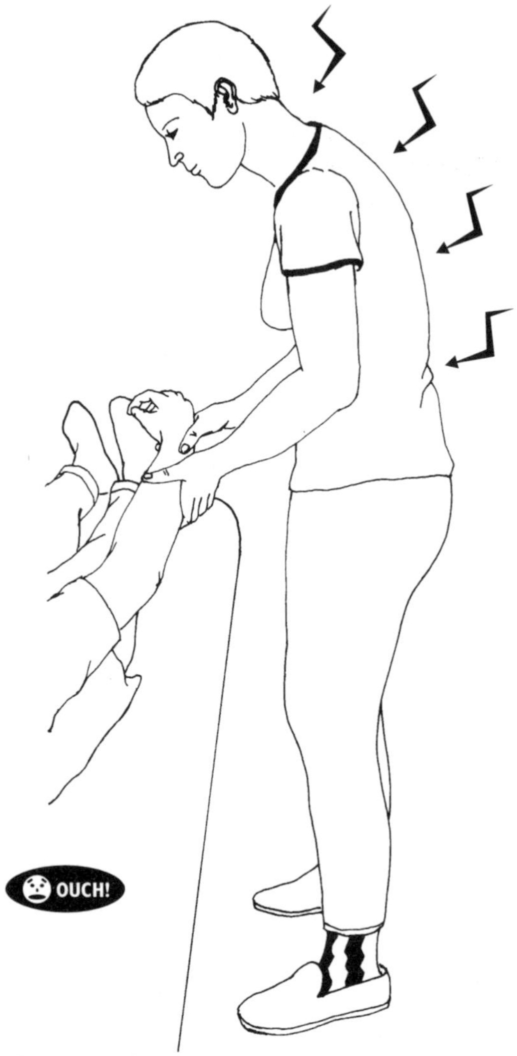

figure 4.14

Imagine that you have a hook at the top of your head, gently lengthening your spine. This will help increase your vertical alignment. Remember, don't think of yourself as standing "straight", because this concept does not allow your spine to maintain its natural curves. Rather, think of the hook as encouraging your spine to remain vertical but also allowing it to keep its natural curvature. (4.15)

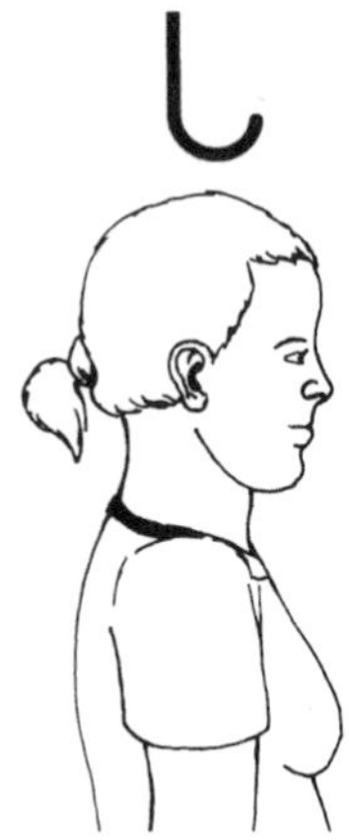

figure 4.15

TRY THIS

Remaining vertical while standing: Part 1

 Stand with your feet under your pelvis. Distribute your weight equally between both feet.

ACTION Begin to slouch or slump your upper body forward. This may feel as if you are forming a convex shape with your spine. **(4.16)**

■ Notice how this position affects your balance and skeletal support.
✔ *Are you able to maintain contact with the tripods of your feet?*
✔ *Is your weight more toward the balls of your feet?*
✔ *Your heels?*

■ Sense how the muscles in your upper body respond to this position.
✔ *Is there muscular effort in your neck?*
✔ *Your upper back?*
✔ *Your lower back?*

■ Sense how the muscles in your lower body respond to this position.
✔ *Is there muscular effort in your thighs?*
✔ *In your calves?*
✔ *In your feet?*
✔ *Can you breathe comfortably in this position?*

ACTION Remain in this position and lift your arms.

■ Sense how lifting your arms in this position affects the muscles in your upper body.
✔ *Is there muscular effort in your neck?*
✔ *Your upper back?*
✔ *How do your shoulders feel as you lift your arms?*
✔ *Is it comfortable to lift your arms from this position?*

STOP *Take a break before continuing with Part 2 on the next page.*

figure 4.16

78

Remaining vertical while standing: Part 2

 Stand with your feet under your pelvis.
Distribute your weight equally between both feet.

ACTION Begin to hyperextend your back so it feels as if
you are forming a concave shape with your spine. **(4.17)**

■ Notice how this position affects your balance and
skeletal support.
   ✔ *Are you able to maintain contact with the tripods of
   your feet?*
   ✔ *Is your weight more toward the balls of your feet?*
   ✔ *Your heels?*

■ Sense how the muscles in your upper body respond
to this position.
   ✔ *Is there muscular tension in your neck?*
   ✔ *In your upper back?*
   ✔ *In your lower back?*

■ Sense how the muscles in your lower body respond to
this position.
   ✔ *Is there muscular tension in your thighs?*
   ✔ *In your calves?*
   ✔ *In your feet?*
   ✔ *Can you breathe comfortably in this position?*

ACTION Remain in this position and lift your arms.

■ Sense how lifting your arms in this position affects the
muscles in your upper body.
   ✔ *Is there muscular effort in your neck?*
   ✔ *In your upper back?*
   ✔ *How do your shoulders feel as you lift your arms?*
   ✔ *Is it comfortable to lift your arms from this position?*

REST

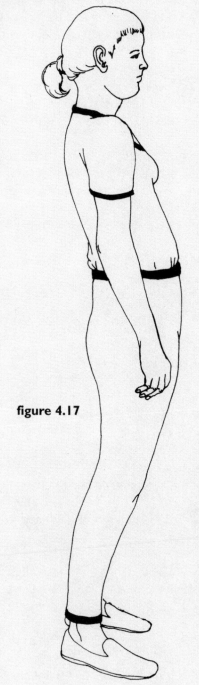

figure 4.17

Stand once again with your feet under your pelvis and distribute your weight equally between both feet.

ACTION Begin to slowly side bend your upper body to the right. Pause for a moment, and then slowly side bend to the left. **(4.18)**

■ Notice what happens to your balance and skeletal support in each position.
✔ *Are you able to maintain contact with the tripods of your feet?*

■ Sense how the muscles in your upper body respond to each position.
✔ *Do they create muscular tension in your neck?*
✔ *In your upper back?*
✔ *In your lower back?*

■ Sense how the muscles in your lower body respond to each position.
✔ *Do they create muscular tension in your thighs?*
✔ *In your calves?*
✔ *In your feet?*
✔ *Can you breathe comfortably in each position?*

STOP *Take a break before continuing with Part 3 on the next page.*

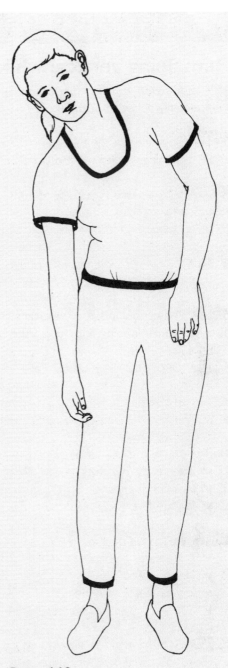

figure 4.18

80

Remaining vertical while standing: Part 3

Stand with your feet under your pelvis, distributing your weight equally between both feet.

ACTION Begin to imagine that you have a hook that is gently attached and lengthening you from the top, center point of your head. **(4.19)**

ACTION Slowly move your upper body a tiny bit forward, backward and side to side. Find for yourself a place where you sense your upper body to be vertically aligned over your pelvis, legs and feet.

■ Notice how your weight is distributed over your feet as you move your upper body in these different directions.

TIP When you find your vertical alignment, chances are you will also feel yourself being fully supported by the tripods of your feet.

ACTION Once you find your vertical alignment, stand for a moment.

■ Notice how this position affects your balance and skeletal support.
✔ *Are you able to maintain contact with the tripods of your feet?*

■ Sense how the muscles in your upper body respond to this position.
✔ *Is there less muscular tension in your neck?*
✔ *In your upper back?*
✔ *In your lower back?*

■ Sense how the muscles in your lower body respond to this position.
✔ *Is there less muscular tension in your thighs?*
✔ *In your calves?*
✔ *In your feet?*
✔ *Can you breathe comfortably in this position?*

ACTION Remain vertically aligned and lift your arms.

■ Sense how lifting your arms in this position affects the muscles in your upper body.
✔ *Is there less muscular effort in your neck?*
✔ *In your upper back?*
✔ *How do your shoulders feel as you lift your arms?*
✔ *Is it comfortable to lift your arms from this position?*

Continue to stand vertically aligned and just enjoy the feeling!

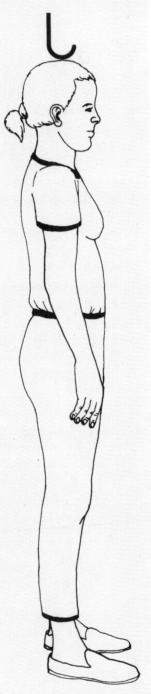

figure 4.19

Where's Your Head?

Your head weighs about 15 pounds. That's a lot of weight to carry around. Not only does your head weigh the same amount as a big bag of potatoes, its center of gravity is in front of your spine. (4.20)

Because your head's center of gravity is in front of your spine, the muscles at the back of your neck, must always be in contraction, to hold your head from falling over. Balancing your head over your shoulders when standing can decrease much of the muscular tension felt in your neck caused from, for example, a forward head posture. Knowing this, it is easy to understand why so many manual therapists express frustration with the amount of stress and fatigue they experience in their necks while standing.

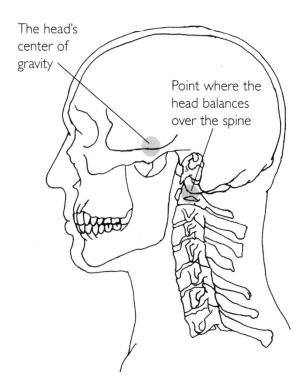

The head's center of gravity

Point where the head balances over the spine

figure 4.20

TRY THIS

Try to sense where your head is in relationship to the rest of your body by putting a book on top of your head.

Sometimes it is hard to sense where your head is in space. Why? Well, your head, unlike the rest of your body, has nothing pushing down on the top of it. The rest of the bones in your body have other bones pressing on them, like the book on your head. This gives them a kind of feedback, telling them where they are in relationship to each other and where they are in space.

PREVENTION TIP

Standing is a very dynamic part of your body mechanics, no matter from which standing positions you choose to work. Don't find yourself standing in one fixed position throughout your treatments. Move around and explore different standing options, keeping the points of this chapter in mind.

82

Do it As Homework

Stand vertically aligned and bring your attention to your head.

■ Sense how and where you hold your head.
✔ *Do you hold your head with your chin up or down?*
✔ *Do you hold your head with your chin turned toward your right or left shoulder?*
✔ *Is your head tilted toward your right or left shoulder?*

ACTION Imagine that your head is a helium balloon and is floating at the top of your spine.

TIP Don't increase your muscular effort to do this. Let your muscles relax and imagine that your head is floating without effort.

ACTION Begin to slowly move your head up and down, as if you are nodding "yes." **(4.21)** Allow your eyes to follow the movement of your head and make very slow and small movements.

■ Sense from which part of your spine you make this movement.
✔ *Do you make this movement from the bottom or the top of your cervical spine?*
✔ *How high up on your cervical spine can you make this movement?*

TIP You may need to make smaller movements in order to feel the movement coming from the top of your cervical spine.

ACTION Continue to slowly nod "yes." Make this movement until you find a place somewhere between up and down where you sense your head to be balanced over your shoulders and cervical spine.

ACTION Look around and sense how your head moves.
✔ *Does your head feel lighter, as if it is floating above your shoulders?*

REST

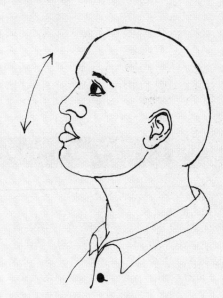

figure 4.21

Once again, stand vertically aligned. Bring your attention to your head and again imagine that your head is a helium balloon, sitting at the very top of your spine.

ACTION Begin to turn your head right and left, as if you are making the movement of "no." Make small and slow movements. **(4.22)**

■ Sense from which part of your spine you make this movement.
  ✔ *Do you make this movement from the bottom or the top of your cervical spine?*
  ✔ *How high up on your cervical spine can you make this movement?*

ACTION Continue to make these movements until you find a place somewhere between right and left, where you sense your head to be balanced and centered over your shoulders and spine.
  ✔ *Is your head sitting at the top of your spine without muscular effort?*

REST

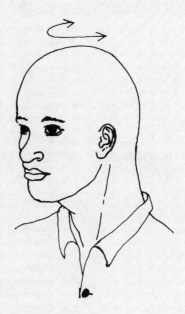

figure 4.22

ACTION Again imagining your head as a helium balloon, begin to tilt your head toward your right shoulder and then tilt your head toward your left shoulder. Be sure to make slow and small movements. **(4.23)**

ACTION Continue this movement until you can find a place between your right and left shoulders where your head is balanced over your shoulders and sitting on the top of your spine.

REST

 Stand and sense where your head is now.
  ✔ *Does it sit on the top of your spine with less muscular effort?*

figure 4.23

ACTION Look around and sense your head as it moves.

Experiencing the connection between your head and spine will help you to be more aware of where your head is while standing. Resting the weight of your head through your spine will also increase your skeletal support in standing.

CHECK THIS OUT

Why do cats always land on their feet? Cats have a fine sense of balance and body position. When a cat begins to fall, it is immediately able to sense its position in space. Then to get its feet pointed toward the ground, it makes a series of twisting motions. And presto! It lands on its feet! This gives cats a real advantage in life because they can use their limbs to cushion a fall, putting them in a position to run, jump, or move in any direction once they land.

STANDING SUMMARY

1

Standing still.
While you are standing and working, you are free to move and not expected to remain in a fixed or static position.

2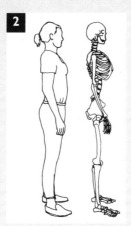

Standing "straight".
Trying to stand "straight" takes your skeleton out of its natural alignment and requires your muscles to work very hard to hold a "straight" posture.

3

Standing on your tripods.
When standing with your body's weight evenly placed through your foot's natural tripod, your entire skeleton has a solid base on which to stand.

4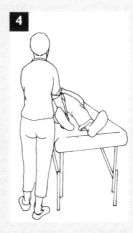

The direction of your feet.
Direct both of your feet forward, and in the direction of your work, to decrease the risk of strain to your lower back, knees and ankles.

5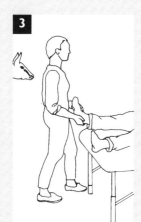

Remain Vertical.
When standing, keep your upper body vertical and balanced over your pelvis, legs and feet. When vertical, your center of gravity remains over your base of support.

6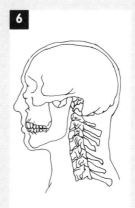

Keep your head over your shoulders.
Balance your head over your shoulders to decrease much of the muscular tension felt in your neck.

Notes:

FYI

CHECK THIS OUT

Mountain goats are great climbers. Why? They are what you call "sure-footed". The shape of their hooves are such that they will not lose balance, slip or fall when climbing.

CHAPTER 5

Sitting

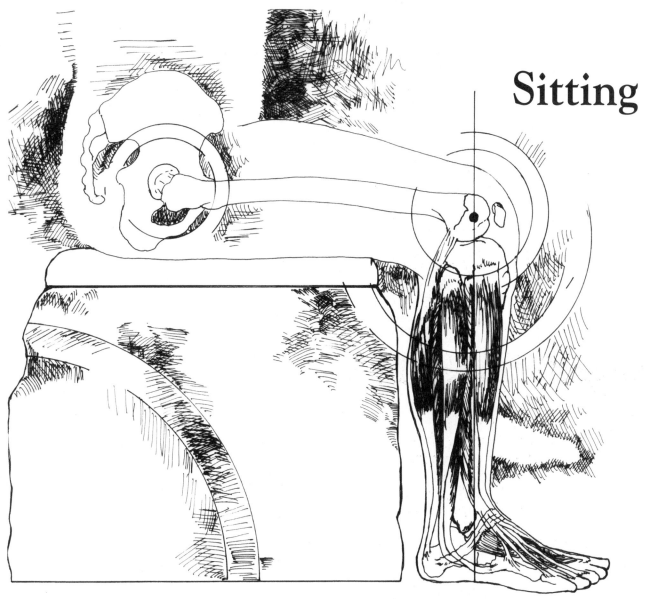

INTRODUCTION

Because our bodies are structured for a bi-pedal life, sitting comfortably often presents a problem. The pelvis, which is oddly shaped and primarily functional for walking, does not provide a very comfortable platform for the body to be in a sitting position.

We will explore three major concepts that will lead you to experience comfortable and effective sitting. First, you will learn how to support yourself with your pelvis, upper posterior thighs and feet. Second, you will learn how to find vertical balance between your pelvis and upper body. And third, you will learn how to balance your head over your shoulders. You will also learn about knee height and leg width, plus a few other tips that will lead to comfortable and effective sitting.

THE HABITS OF EVERYDAY LIFE...

How do you normally sit throughout the day?

◆ How do you sit to watch TV?

◆ Drive your car?

◆ Eat dinner?

◆ Use your computer?

◆ What is your most comfortable sitting position?

◆ Are you generally comfortable sitting?

◆ Do you sit with your legs crossed?

◆ Do you sit with one leg tucked underneath you?

◆ Do you sit with your feet in contact with the floor?

AS A MANUAL THERAPIST. . .

How do you sit?

◆ Do you sit with both of your feet on the floor?

◆ Do you sit with one leg crossed over the other?

◆ Do you sit using back support?

◆ Do you sit with your legs stretched out
in front of you?

◆ Do you sit on the edge of your chair?

◆ Do you tend to slouch forward?
◆ Do you arch your low back?

◆ Do you lead with your head while sitting?

◆ Do you sit frequently throughout your treatments?

◆ Are you comfortable sitting?

The Hip Bone's Connected to the Thigh Bone...

The bones of your pelvis form a kind of bowl that offers support for sitting. The ischial tuberosities, commonly called the "sit-bones", are the lowest part of the bowl, where the weight of your upper body ideally rests. The sit-bones usually get all of the attention when it comes to sitting. (5.1)

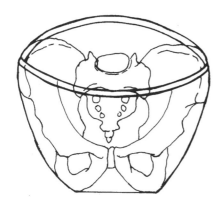

figure 5.1

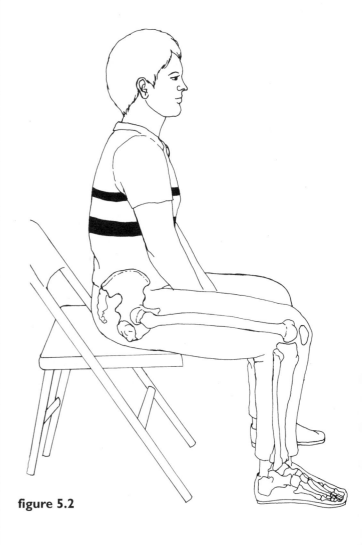

figure 5.2

PREVENTION TIP

The hip joints connect the upper body to the lower when sitting. Often the lumbar spine is used as the hinging point, causing stress and pain to the lower back. When finding vertical balance for your upper body, hinge from your hip joints and save your low back from discomfort.

Your upper posterior thighs also play an important role in supporting your upper body. Your thighs share the weight of your upper body and stabilize it from falling off the chair. After all, if not for your thighs, your pelvis wouldn't have a chance of holding you up.

Your feet do the rest. They support the weight of your legs and give your upper body additional support. Together, your pelvis, thighs and feet give you a wonderful base of support when sitting. (5.2)

TRY THIS

Palpating your "sit-bones"

Palpating your sit-bones will help you to feel them more clearly when sitting.

 Stand and place one foot on a chair.

ACTION Place your fingers in the middle of your buttock. There you will find what feels like the point of a bone. This is your ischial tuberosity. **(5.3)**

Trace the surface of the ischial tuberosity several times until you gain a clear idea of its actual shape and size.

Once you have palpated one side, switch legs and palpate the other side.

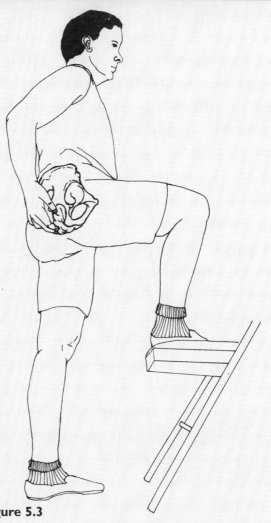

figure 5.3

PREVENTION TIP

If needed, sit with a rolled up towel underneath your tailbone. This can help relieve pressure placed on the sacrum. Make sure the towel is underneath your tailbone and not your sit-bones. Your sit-bones should continue to rest on your chair.

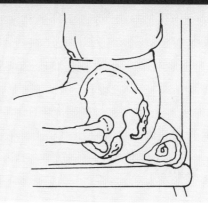

94

Making the pelvis, thigh, and foot connection: Part I

Sit on a firm surface that is flat and level. Place your ankles under your knees, and allow your knees and feet to be as wide as your pelvis. **(5.4)**

✔ *What part of your pelvis are you sitting on right now?*

ACTION Begin to tilt your pelvis forward and backward.

✔ *Can you feel yourself rolling over your ischial tuberosities?*

TIP Your ischial tuberosities are shaped like the base of a rocking chair that can easily rock your pelvis forward and backward. **(5.5)**

ACTION Continue to roll your pelvis back and forth over these bones so you become very aware of where they are.

REST

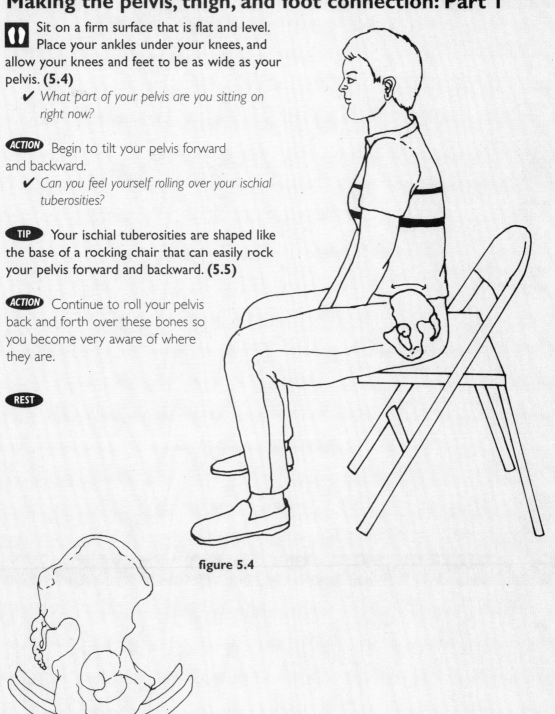

figure 5.4

figure 5.5

Again, place your ankles under your knees, and allow your knees and feet to be as wide as your pelvis.

ACTION Roll your pelvis back and remain on the posterior edge of your ischial tuberosities for a moment. **(5.6)**
✔ *How does this sitting position affect your low back?*
✔ *On what part of your pelvis is your weight resting?*
✔ *How much weight is on your femurs?*
✔ *Does this sitting position feel like a familiar one to you?*

ACTION Slowly begin to tilt your pelvis forward so you begin to transfer some of your weight onto your thighs.

TIP You should now be resting your weight on the base of your pelvis and your thighs. **(5.7)**
✔ *How does this sitting position affect your low back?*
✔ *On what part of your pelvis is your weight resting?*
✔ *How much weight is on your thighs?*

TIP Most of your weight while sitting is supported by your pelvis. Realize that your thighs also support your weight. Together, they take the pressure and strain off of your low back and allow it to remain in a neutral position.

STOP *Take a break before continuing with Part 2 on the next page.*

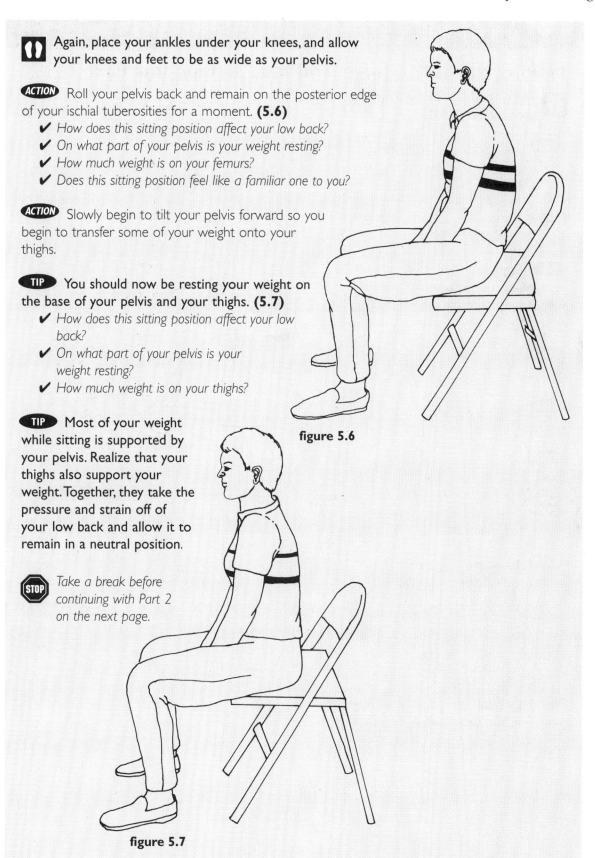

figure 5.6

figure 5.7

TRY THIS

Making the pelvis, thigh, and foot connection: Part 2

👣 Once again sit with your weight over your pelvis and thighs. **(5.8)**

ACTION Now bring your attention to your feet.
✔ *At this moment, how are they supporting your legs?*

ACTION Arrange your feet so your ankles are aligned underneath your knees.

TIP Bring your ankles underneath your knees to increase your skeletal support and decrease the muscular effort in your legs.

ACTION Lift and lower your heels a few times.

ACTION Lift and lower the balls of your feet a few times.

ACTION Lift and lower the outer edges a few times.

ACTION And finally, lift and lower the inner edges a few times.

ACTION Now bring your feet in full contact with the floor and allow them to fully support your legs.

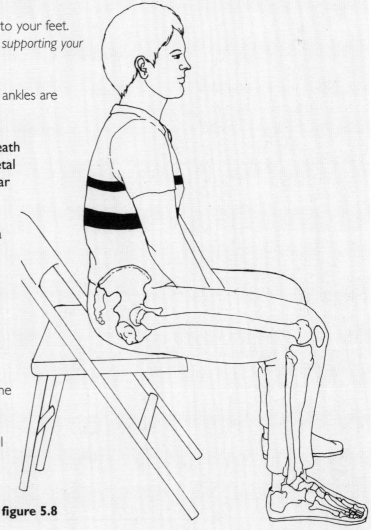

figure 5.8

Feel the support and stability you now have with your weight on your pelvis, thighs and feet.

Vertical Balance

When sitting, your center of gravity is lowered and your base of support is larger then when standing. Sitting on a large base of support instead of standing on a small one may give you the impression that vertical alignment is not as important while sitting. Nothing could be farther from the truth. In fact it is just as important to maintain vertical balance while sitting.

Sitting is believed to cause the majority of low back pain in this country. This is due to the fact that people tend to sit in a slouched or slumped position for hours at a time, increasing the pressure on their low back. In contrast to balanced vertical sitting, a slouched sitting position stresses the discs and posterior ligaments of the back. Overall there is an increased risk of pain and stress to the lower and upper back as well as the neck. (5.9)

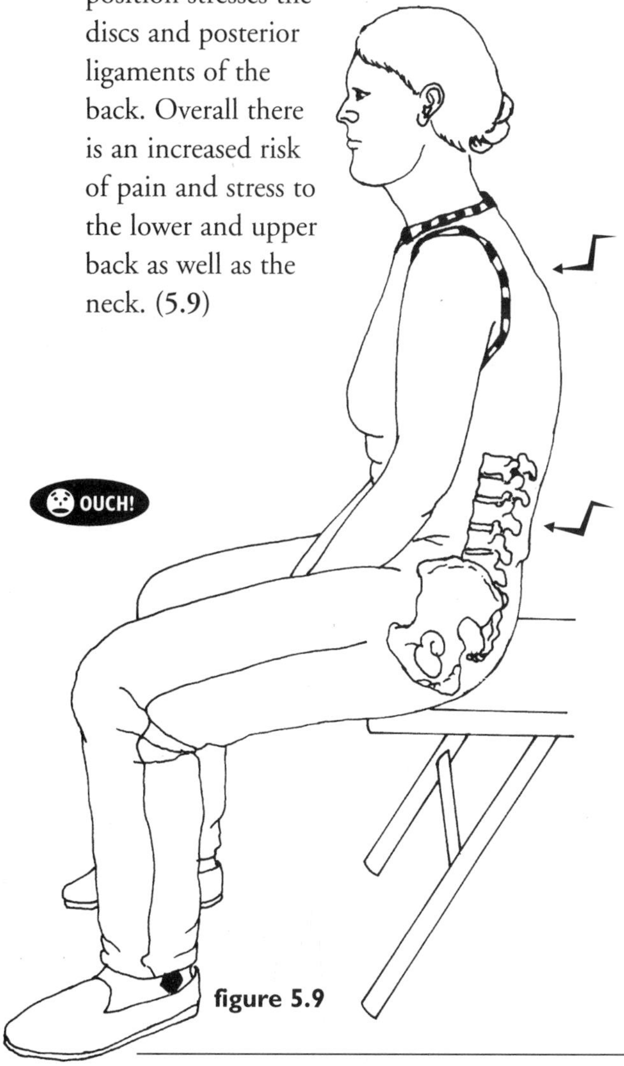

OUCH!

figure 5.9

Sitting with your upper body vertically balanced over your pelvis allows your low back to maintain its lumbar curve, decreasing the pressure from flexion. Not only does it allow your low back to relax, it also allows your entire back to keep its natural curves, reducing excessive muscular effort. When your muscular tension is reduced, your shoulders, arms and hands gain flexibility and movement to facilitate your manual therapy. (5.10)

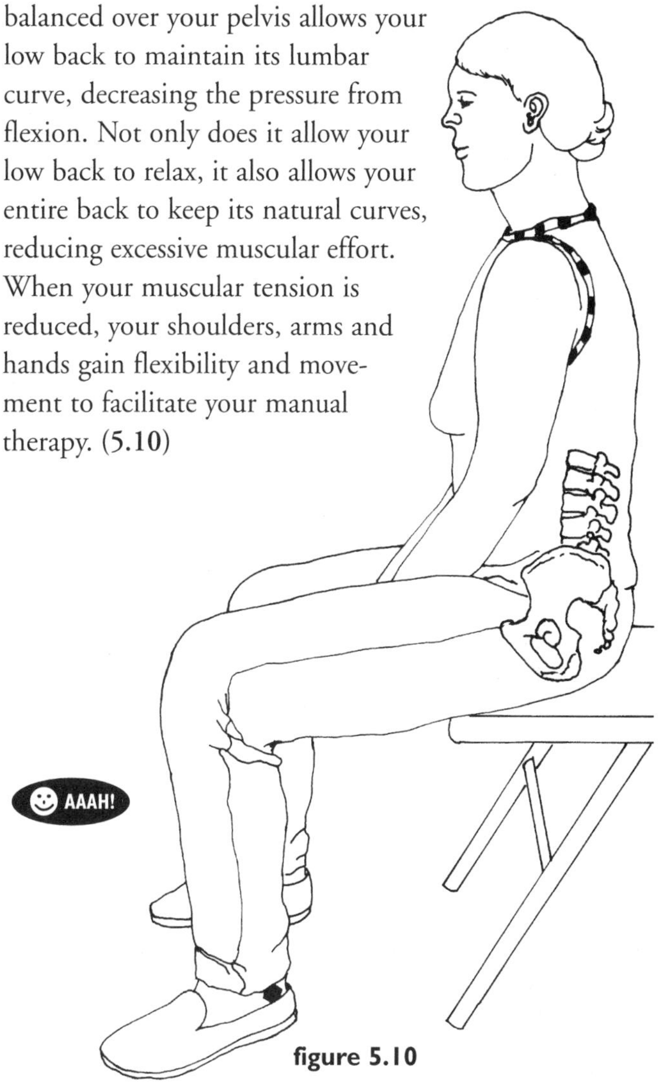

AAAH!

figure 5.10

CHECK THIS OUT

Studies have shown that unbalanced sitting postures are major factors in thoracic outlet syndrome. It is also believed that thoracic outlet syndrome is directly related to carpal tunnel syndrome.

TRY THIS

Balancing your pelvis and upper body: Part I

👣 Sit on a chair with a flat and level surface. Place your ankles under your knees.

ACTION Begin to roll or tilt your pelvis back. Let your back, shoulders and head follow the movement of your pelvis. **(5.11)**

TIP This movement is very similar to slumping or slouching down in a chair. In this position, your upper body is collapsing so that your spine has no chance to support you.

■ Notice that when you tilt your pelvis back, your head moves forward, forcing your upper body to flex or slouch.
　✔ *Can you sense that your upper body and pelvis are not vertically balanced in this position?*
　✔ *Can you sense that your lumbar curve has decreased?*

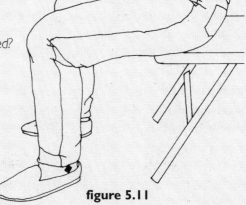

figure 5.11

■ Notice how the muscles in your back respond to this slouched position.
　✔ *Can you feel muscular tension in your low back?*
　✔ *In your mid and upper back?*
　✔ *Can you breathe comfortably from this slouched position?*

TIP In this position, your diaphragm is severely restricted, making it hard for you to breathe normally.

ACTION Raise your arms from this position. **(5.12)**
　✔ *Do you feel a sense of effort while raising your arms?*
　✔ *Where do you feel the effort?*
　✔ *In your shoulders? Back? Neck?*

■ Notice if this is a position from which you tend to work.

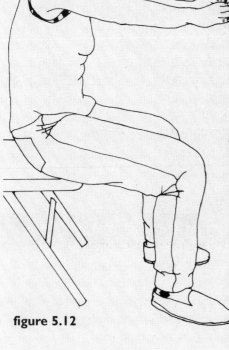

figure 5.12

REST

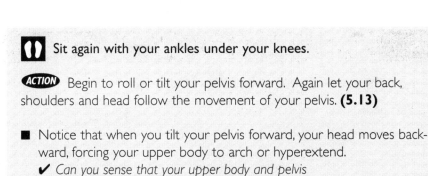 Sit again with your ankles under your knees.

ACTION Begin to roll or tilt your pelvis forward. Again let your back, shoulders and head follow the movement of your pelvis. **(5.13)**

■ Notice that when you tilt your pelvis forward, your head moves backward, forcing your upper body to arch or hyperextend.
  ✔ *Can you sense that your upper body and pelvis are not vertically balanced in this position?*
  ✔ *Can you sense that your lumbar curve has increased?*

■ Notice how the muscles in your back respond to this arched position.
  ✔ *Can you feel muscular tension in your low back?*
  ✔ *In your mid and upper back?*
  ✔ *Can you feel how your shoulders are forced back?*

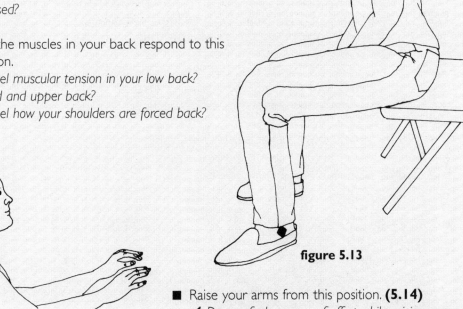

figure 5.13

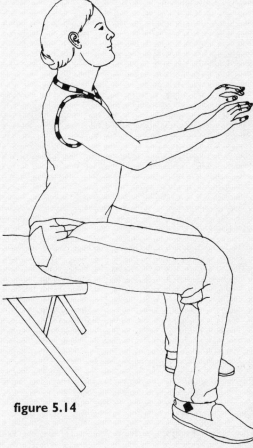

■ Raise your arms from this position. **(5.14)**
  ✔ *Do you feel a sense of effort while raising your arms?*
  ✔ *Where do you feel the effort?*
  ✔ *In your shoulders?*
  ✔ *In your back?*
  ✔ *In your neck?*
  ✔ *Can you breathe comfortably from this arched position?*

■ Notice if this is a position from which you tend to work.

STOP *Take a break before continuing with Part 2 on the next page.*

figure 5.14

100

Balancing your pelvis and upper body: Part 2

👣 Sit again with your ankles under your knees.

ACTION Roll or tilt your pelvis to a place where your pelvis feels as if it is resting on its base and where your weight is being supported by your sit-bones and upper posterior thighs. **(5.15)**

TIP This place will be a neutral position, somewhere between the backward and forward movements previously made.

ACTION Allow your back, shoulder and head to follow the tilting movement of your pelvis so that you eventually find a place where your upper body feels vertically balanced over your pelvis.

- ✔ *Can you sense that your upper body and pelvis are vertically balanced in this new position?*
- ✔ *Can you sense that your lumbar curve is in a neutral position?*

■ Notice how the muscles in your back respond to being vertically aligned with your pelvis.

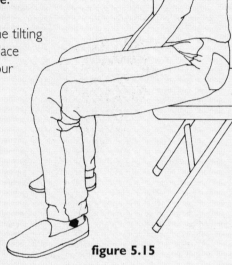

figure 5.15

- ✔ *Can you sense less muscular effort in your low back?*
- ✔ *In your mid and upper back?*
- ✔ *Are your shoulders in a more comfortable place?*

■ Notice your breathing.
- ✔ *Can you breathe more comfortably now?*

■ Raise your arms from this position. **(5.16)**
- ✔ *Do you feel a sense of ease while raising your arms?*
- ✔ *Where do you feel the ease?*
- ✔ *In your shoulders? Back? Neck?*

ACTION Lower your arms and continue to sit for a few minutes. **Enjoy the feeling of sitting in a effective and comfortable manner!**

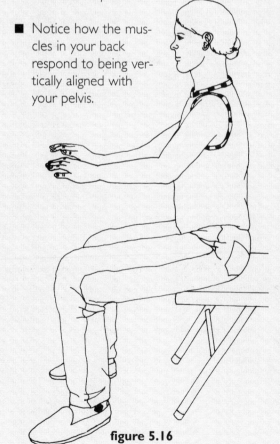

figure 5.16

Where's Your Head?— The Sequel...

A goal of most manual therapists is to reduce the amount of neck tension felt while sitting. Even if you are sitting in a vertically balanced manner, if your head is not part of the balance, chances are you will suffer from neck pain and stress. (5.17) Because the head weighs about 15 pounds, it is not easy to always have it under control. However, you can prevent neck stress and pain by focusing on balancing your head over your shoulders. (5.18)

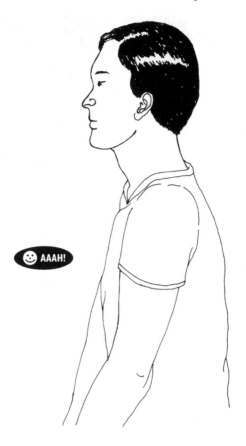

figure 5.18

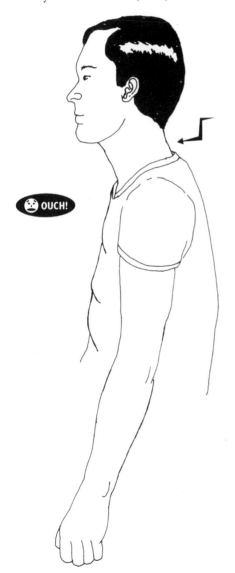

figure 5.17

PREVENTION TIP

Sitting vertically does not mean you should sit up "straight." As in standing, the notion of sitting up straight creates a struggle for the muscles to hold a certain posture. Remember, there is not one straight bone in the body. Therefore, sitting up "straight" is going against the natural curves and dynamics of your skeleton.

102

Finding your head: Part 1

Sit with your weight on your sit-bones, posterior thighs and feet. Find the vertical balance between your pelvis and your upper body.

- Notice where your head is.
  - ✔ *Can you feel where your head is in relationship to your shoulders?*

ACTION Put your hand on top of your head and see if it helps you sense where your head is.

ACTION Keep your hand on the top of your head and move your head out in front of your shoulders. **(5.19)**

- Notice how heavy your head feels in this position.
  - ✔ *How does this affect the muscles in your neck?*
  - ✔ *Does it cause tension?*

- Notice how this position affects your upper back and shoulders.
  - ✔ *Does it cause tension in these areas?*

ACTION Leave your head out in front of your shoulders and lift both of your arms. **(5.20)**
  - ✔ *How easy is it to lift your arms?*
  - ✔ *How does lifting your arms in this position affect the muscles in your upper back?*
  - ✔ *How does it affect your chest and your breathing?*

ACTION Slowly look around and notice if it is comfortable to use your eyes.

REST

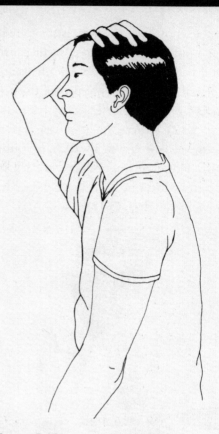

figure 5.19

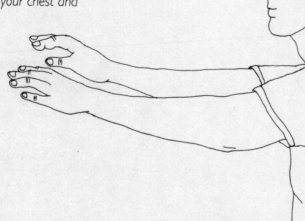

figure 5.20

ACTION Once again place your hand on the top of your head.

ACTION Now move your head back so that you decrease your cervical curve. **(5.21)**

TIP This position might feel as if you are trying to tuck in your chin.

■ Notice how your head feels in this position.
   ✔ *How does it affect the muscles in your neck?*
   ✔ *Does it cause tension?*

■ Notice how this position affects your back and shoulders.
   ✔ *Does it cause tension?*

ACTION Leave your head in this position and lift both of your arms. **(5.22)**
   ✔ *How does lifting your arms with your head in this position affect your neck?*
   ✔ *Your upper back?*
   ✔ *Your chest?*
   ✔ *Your breathing?*

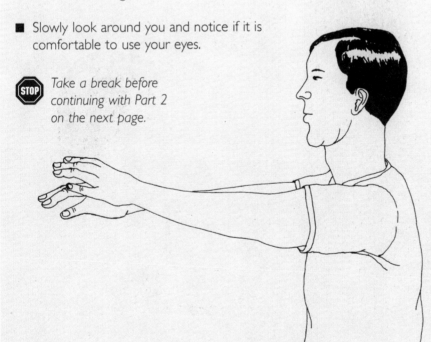

figure 5.21

■ Slowly look around you and notice if it is comfortable to use your eyes.

STOP *Take a break before continuing with Part 2 on the next page.*

figure 5.22

104

Finding your head: Part 2

👣 Sit again vertically balanced.

ACTION Return your hand to the top of your head.

ACTION Move your head until you find a place where your head feels balanced over your shoulders. **(5.23)**

TIP This place will be somewhere between the forward and backward positions you just experienced.

ACTION When you feel as if you have found a balanced place for your head, remove your hand. Nod your head "yes" a few times and "no" a few times.

TIP If your head is balanced, you will feel very little muscular tension in your neck while moving your head up, down and side to side.

■ Notice if your upper back feels more comfortable with your head balanced over your shoulders.

ACTION Lift your arms again. **(5.24)**
    ✔ *Is it easier than before?*

ACTION Look around and notice if it is easier to use your eyes.

figure 5.23

TIP Sit and work with your head balanced over your shoulders to increase the comfort in your back and shoulders.

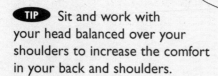

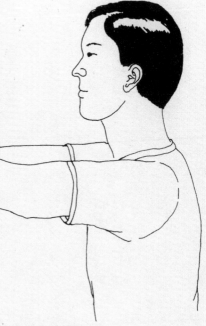

figure 5.24

Knee Height

When sitting it is important that your knees are the same height as your hips.

When your knees are higher then your hips, your pelvis is forced back and the vertical balance of your pelvis and upper body is compromised. (5.25)

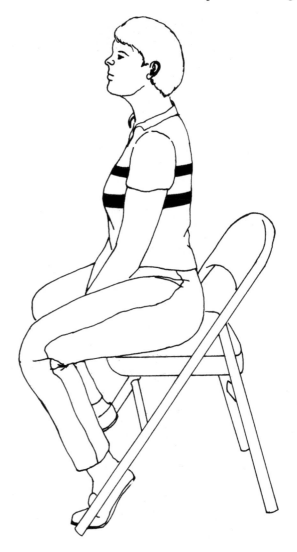

figure 5.26

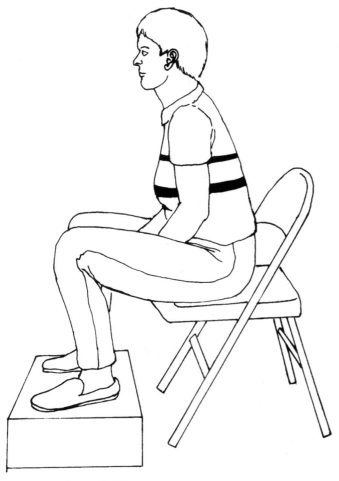

figure 5.25

When your knees are lower then your hips, your pelvis rolls forward and again your vertical balance is compromised. (5.26)

Keeping your knees the height of your hips increases the comfort and effectiveness of your sitting. (5.27)

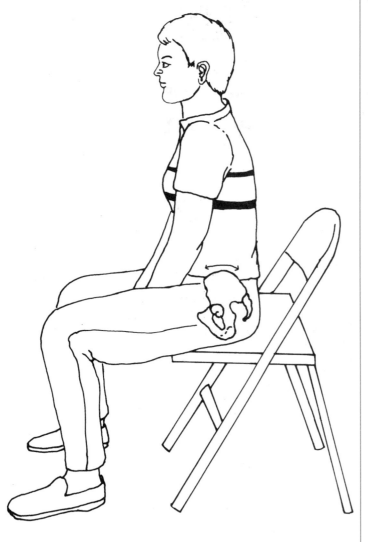

figure 5.27

Muscle tension of the posterior neck is increased by 50 percent when sitting in a slouched or flexed position.

Leg Width

The width of your legs influences the position of your pelvis, which greatly affects your muscular tension, stability and vertical balance. Therefore, holding your legs close together or spreading them wide apart can severely increase the muscular effort and tension in your pelvis (e.g., your gluteals) and even your back (e.g., your erector spinea). Finding the right width for your legs helps maintain your stability and vertical balance, and reduces muscular tension.

PREVENTION TIP

If you sit with your legs wide to accommodate the structure of your table, keep your back in a neutral position. Sitting with the legs wide creates a tendency to hyperextend the spine. This can cause general discomfort and pain to your body, if held for long periods of time.

Keep your back in a neutral position, as much as possible, to help reduce the stress. You may even find that it allows you to sit more comfortably when straddling your table.

TRY THIS

Finding the right width: Part I

 Sit on a flat and level surface with your knees at hip
height. Sit with your legs and feet very close together,
so your thighs and ankles are touching each other. **(5.28)**

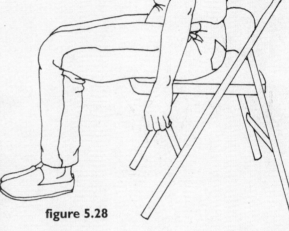

■ Notice how sitting with your legs close together affects your
pelvis and upper body.
✔ *Do you feel effort in your pelvis?*
✔ *Your low back?*
✔ *Your upper back?*
✔ *Your neck?*

■ Notice the position of your pelvis.
✔ *Is it rolled back or forward?*
✔ *Can you feel the muscles of your legs
working hard to stay together?*
✔ *Does this sitting position create a stable
base of support?*
✔ *Is this a position in
which you usually sit?*

REST

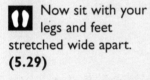 Now sit with your
legs and feet
stretched wide apart.
(5.29)

■ Notice how sitting in this position affects
your pelvis and upper body.
✔ *Do you feel effort in your pelvis?*
✔ *Your low back?*
✔ *Your upper back?*
✔ *Your neck?*

■ Notice the position of your pelvis.
✔ *Is it rolled back or forward?*
✔ *Does this sitting position create a stable
base of support?*
✔ *Is this a position in which you
usually sit?*

STOP *Take a break before continuing with
Part 2 on the next page.*

figure 5.28

figure 5.29

108

Finding the right width: Part 2

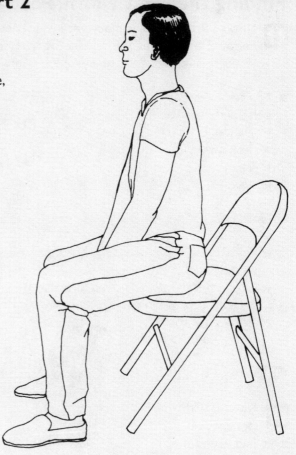

Now sit with your legs as wide as your pelvis. **(5.30)**

TIP Depending on how long your legs are, you may find it more comfortable to sit with your legs a bit wider then your pelvis.

■ Notice how your pelvis and upper body respond to this position.
✔ *Do you feel less effort in your pelvis?*
✔ *Your low back?*
✔ *Your upper back?*
✔ *Your neck?*

■ Sense how you are now creating a stable base of support.
✔ *Can you easily maintain your vertical balance?*

figure 5.30

SITTING SUMMARY

1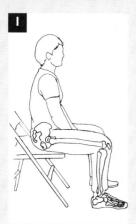

Base of support.
Together, your pelvis, thighs and feet give you a wonderful base of support when sitting. Most of your weight is supported by your pelvis, but your thighs and feet also support your weight.

2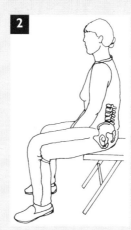

Vertical balance.
Sit with your upper body vertically balanced over your pelvis to allow your entire back to keep its natural curves, reducing excessive muscular effort.

3

Where's your head?
Even if you are sitting in a vertically balanced manner, if your head is not part of the balance, chances are you will suffer from neck pain and stress.

4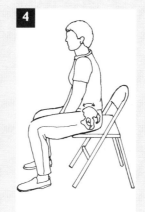

Knee height.
Make sure your knees are the same height as your hips. This gives the bones of your legs the opportunity to maximally support you.

5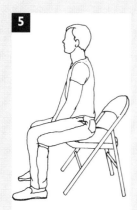

Leg width.
The width of your legs influences the position of your pelvis, which greatly affects your muscular tension, stability and vertical balance.

Notes:

PREVENTION TIP

Don't sit in static positions for long periods of time. Shift you weight around your base of support by bending from your hip joints. Bending from your hip joints will allow you to shift your weight around your pelvis and thighs without compromising your vertical balance. (Bending, page 124)

CHAPTER 6

Bending

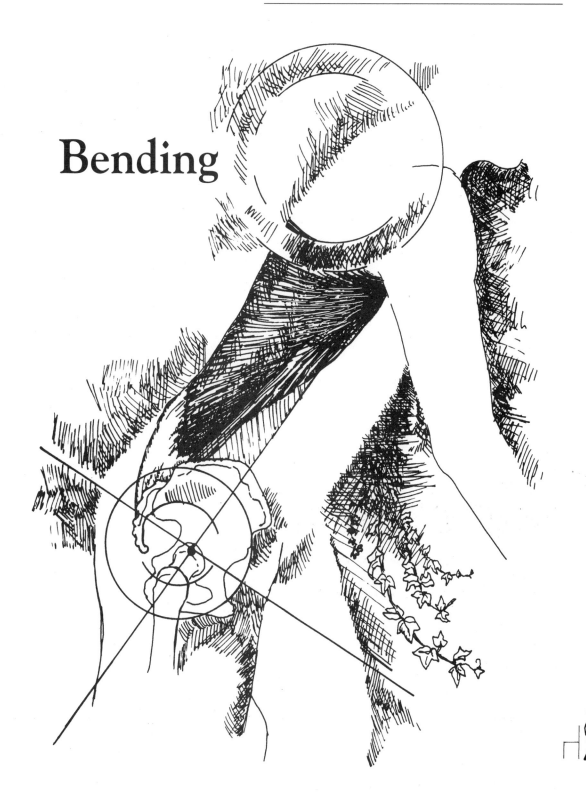

INTRODUCTION

*A*s a manual therapist, your focus is primarily downward, meaning your work is lower then your head. Consequently, your body must flex and bend to accommodate your downward focus and carry out your specific manual therapy. While bending is unavoidable, you can avoid discomfort caused from bending.

Bending from your hip joints, counterbalancing your head and pelvis and integrating the support of your legs and feet, will be our focus to help keep your body, especially your back, safe from discomfort and injury while bending.

THE HABITS OF EVERYDAY LIFE...

How do you normally bend throughout the day?

◆ Over your kitchen sink?

◆ To pick up something you've dropped?

◆ To pet your cat or dog?

◆ Do you bend from somewhere in your upper back?

◆ From the middle of your back?

◆ Your low back?

◆ Do you avoid bending?

◆ If so, how do you compensate?

AS A MANUAL THERAPIST. . .

How do you bend?

◆ Do you bend using your neck?

◆ Your upper back?

◆ Your low back?

◆ Your mid back?

◆ Your knees?

◆ Your hip joints?

◆ Are you aware of when and how often you bend?

◆ Are you comfortable bending?

Your "False Hip Joints"

Constant bending is often the cause of back pain for manual therapists. It is therefore important to understand the difference between bending from your "true hip joints" and bending from your "false hip joints."

Your false hip joints are the areas in your back from where you may habitually bend. Typically, these areas are the upper, middle or low back. (**6.1**) When you constantly bend from your back, the muscles of your back are lengthened, strained and fatigued, and ultimately weakened, which can cause chronic pain. The integrity of your spine is also compromised, putting it at risk for injury.

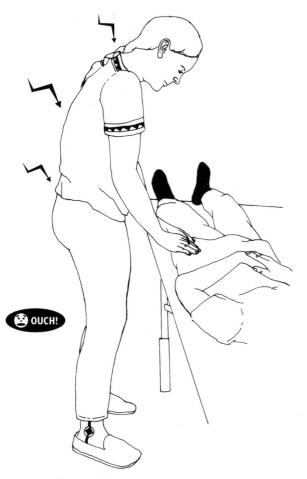

😖 OUCH!

figure 6.1

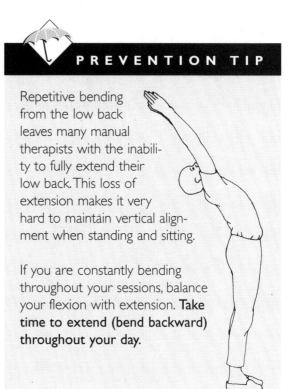

PREVENTION TIP

Repetitive bending from the low back leaves many manual therapists with the inability to fully extend their low back. This loss of extension makes it very hard to maintain vertical alignment when standing and sitting.

If you are constantly bending throughout your sessions, balance your flexion with extension. **Take time to extend (bend backward) throughout your day.**

TRY THIS

Identifying your false hip joints: Part 1

This lesson will help you recognize from where in your back you tend to bend. During the lesson, identify your "false hip joint" and see if this place corresponds to an area of your back that is typically tired or sore after your treatments.

Bending from your false hip joints:

Stand next to your table, with your feet parallel.

ACTION Reach your hands toward your table, bending forward using your neck, upper back and shoulders. **(6.2)**

TIP This movement may feel like you are slumping or slouching your shoulders.

■ Notice how the muscles of your neck, upper back and shoulders feel as you make this movement.
   ✔ *Can you feel effort in your trapezius muscle?*
   ✔ *Can you easily reach your arms toward your table?*
   ✔ *Does this feel like a familiar place from where you bend?*

REST

ACTION This time, reach your hands toward your table, primarily bending from the middle part of your back. **(6.3)** Your shoulders will also be involved in this movement, but try to initiate the bending from your mid back.

TIP This movement may feel like you are curving your mid back to form the letter "C".

■ Notice how your chest and rib cage feel as you make this movement.
   ✔ *How does bending from your mid back affect your breathing?*
   ✔ *Can you easily reach your arms toward your table?*
   ✔ *Does this feel like a familiar place from where you bend?*

STOP *Take a break before continuing with Part 2 on the next page.*

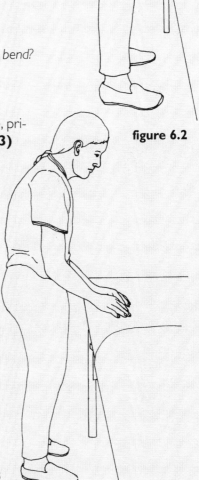

figure 6.2

figure 6.3

Identifying your false hip joints: Part 2

 Again, stand next to your table, with your feet parallel.

ACTION Reach your hands toward your table, this time bending primarily from your low back. **(6.4)** Your upper and mid back will bend as well, but try to initiate the bending from your low back.

TIP This movement may feel as if you are curving your low back to form a concave shape.

■ Notice in what direction your pelvis moves when you bend from your low back.
✔ *Does it tilt forward or backward?*
✔ *How does your low back feel when bending this way?*
✔ *Can you easily reach your arms toward your table?*
✔ *Does this feel like a familiar place from where you bend?*

REST

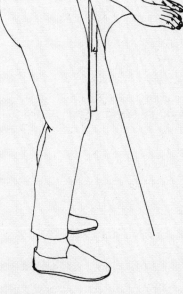

figure 6.4

✔ *Where in your back did you find you most commonly bend?*
✔ *Does this place correspond to a place that is usually fatigued or sore after you work?*

Your True Hip Joints

Your hip joints are the strongest and one of the most stable in your body. They also are the powerful joints that connect your upper and lower body. When you bend from your hip joints, you recruit the strong and powerful muscles of your pelvis and legs. (**6.5**) These muscles along with the strong and stable ball and socket joints of your hips can easily support your weight and facilitate your bending movements. This relieves the muscular effort of your back and decreases spinal stress.

When you hinge or bend from your back, you literally hang the weight of your head and upper body from the bending point of your spine. (**6.6**) (Remember your head weighs approximately 15 pounds!) This requires the thin, long muscles of your back to work extremely hard and places stress on the delicate vertebrae of your spine.

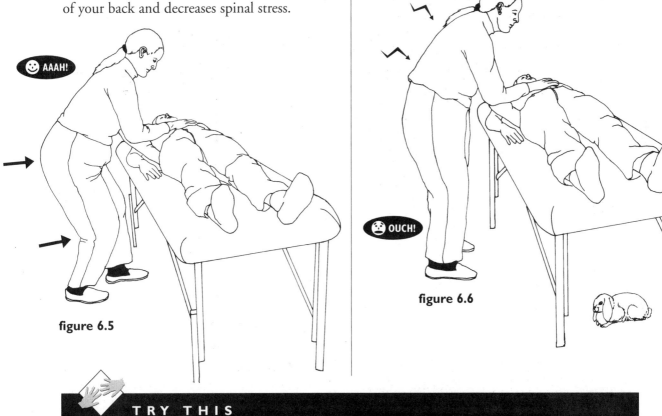

figure 6.5

figure 6.6

TRY THIS

How hip are you?

ACTION Stand and point to where you think your hip joints are with your index finger.

✔ *Where are you pointing?*
✔ *Are you pointing to the outside of your hips—where your greater trochanter is?*
✔ *Are you pointing to the inside of your hips—toward the crease of your pelvis and leg?*

TIP If you are pointing to the inside of your hips, congratulations! You are right! Your hip joints are actually inside, where your pelvis and leg form a crease. (**6.7**)

figure 6.7

120

Bending from you true hip joints

 Stand next to your table, vertically aligned and in a wide, parallel stance.

■ Notice how your back feels when vertically aligned.
  ✔ *Can you imagine bending from your hip joints and maintaining your vertical alignment?*
  ✔ *Is your head balanced over your shoulders?*

ACTION Reach your hands toward your table and bend using your hip joints. **(6.8)** As you bend your upper body forward, bend your knees and move your pelvis backward. Keep your upper, mid and lower back in a neutral position.

TIP It is common to hyperextend or exaggerate the lumbar curve when bending from the hip joints. However, your low back should remain in a neutral position as you bend from your hip joints.

ACTION Reach toward your table several times, bending from your hip joints.

ACTION Each time you bend, become more and more clear that you are bending from your true hip joints and not your false hip joints.

■ Notice the freedom of movement you have in your arms when bending from your hip joints.
  ✔ *Remember how difficult it was to move your arms in the "false hip joint" lesson?*

REST

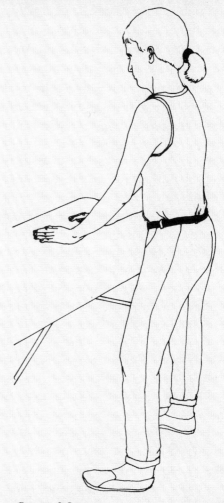

figure 6.8

Stand this time with one foot forward. **(6.9)**

ACTION Reach toward your table, bending from your hip joints. Even though you are standing in a different position, continue to flex both knees and move your pelvis backward.

TIP The tendency in this stance is to only flex the front knee—be sure to flex both.

TIP Bend from your hip joints to allow your back to stay in a neutral position and increase the flexibility and movement in your shoulders, arms and hands.

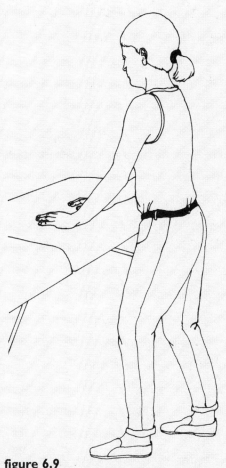

figure 6.9

CHECK THIS OUT

Just how strong are ball and sockets joints? Well, the folks who built Stonehenge knew. Stonehenge is a megalithic monument on Salisbury Plain in Wiltshire, England, where massive horizontal stones were placed on top of massive vertical stones by the use of ball and socket joints. These stones weigh as much as 26 tons! The only thing keeping these huge megaliths in place are their ball and socket joints. Now that's one strong and stable joint!

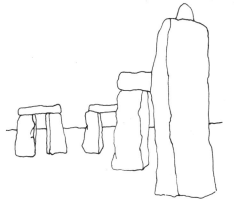

Counterbalance

As you experienced in the last lesson, when bending from your hip joints, it is important to move your pelvis backward. With the movement of your pelvis backward, you are actually counterbalancing the weight of your head and upper body with the weight of your pelvis. Counterbalancing your head and pelvis keeps your center of gravity over your legs and feet. With this kind of balance, you can freely move and bend from your hip joints, leaving your back, shoulders, arms and hands free to facilitate your manual therapy. **(6.10)**

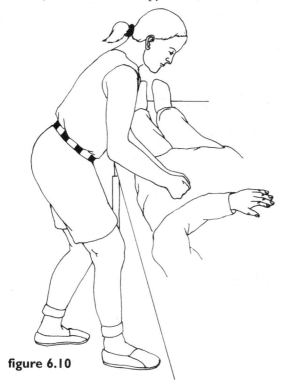

figure 6.10

 PREVENTION TIP

Bending from your true hip joints may seem strange to you. However, if you bend from your hip joints instead of your back, you will decrease your back discomfort and fatigue. This will increase your effectiveness and comfort while bending!

 CHECK THIS OUT

Not only do surfers live to ride the perfect wave, they also know a thing or two about bending and counterbalance. Looking at how a surfer rides his board, you will see he bends from his hip joints and counterbalances the weight of his head and pelvis. Bending from his hip joints allows his upper body to remain flexible and free to move. Counterbalancing allows him to remain stable on his feet and his board.

Knees and Feet While Bending

It is important to place your weight over the tripods of your feet while bending. (Standing, page 71). This helps to increase your stability. It is also important that your knees stay in alignment with your feet. Ideally, your knees should remain over your feet, but when this is not possible, make sure that your knees stay in line with your feet to increase your skeletal support and decrease stress in your knee joints. (6.11)

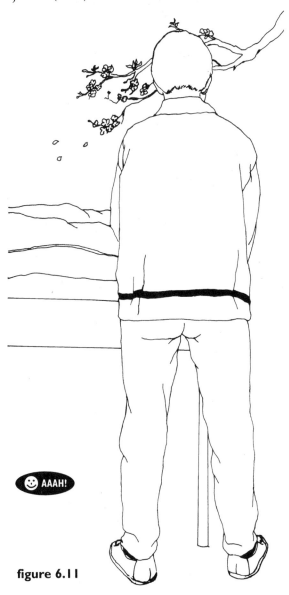

figure 6.11

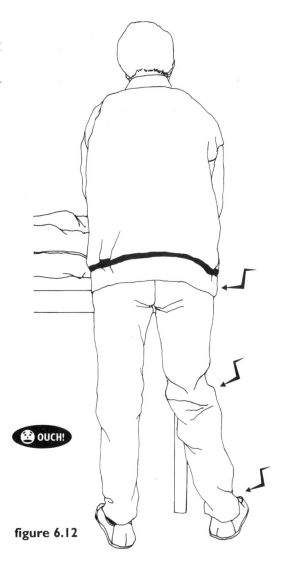

figure 6.12

When your knees are not in alignment with your feet, the muscles, ligaments and tendons of your knees must work very hard to remain stable. (6.12) By keeping your knees in alignment—and when possible—over your feet, you take the strain off your knee joints and allow your skeleton to do the work it was intended to do.

124

Bending While Sitting

When sitting you perform many aspects of your manual therapy, e.g., lifting, pushing and pulling.

Just as you experienced with bending from a standing position, bending from your hip joints and using your legs and feet for support when sitting will decrease the stress and discomfort in your upper body.

Don't let your sitting position fool you. Just because you are sitting does not mean you can bend from your neck or back. You must still use the strong ball and sockets joints of your hips to balance the weight of your upper body's movement forward. (6.13)

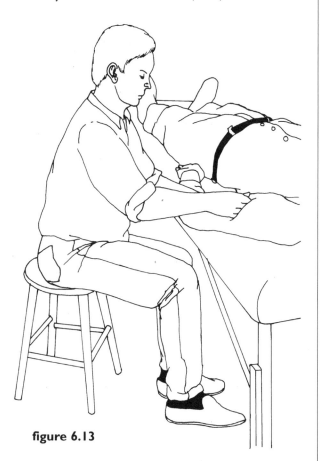

figure 6.13

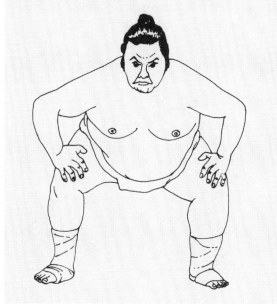

TRY THIS

Bending while sitting: Part I

 Sit on a chair with a flat and level surface. Sit vertically balanced with your weight on your sit-bones, posterior thighs and feet. (Sitting, page 92)

ACTION Begin to bend in a way you would normally bend if you were working with a client.

■ Notice from where you begin to bend.
  ✔ *Do you start the bending from your mid or upper back?*
  ✔ *Your neck?*

ACTION Now imagine there is a table that sits just out in front of your knees. **(6.14)**

ACTION Bend forward to put your nose on the table.

■ Notice how you make this movement.
  ✔ *Did you bend from your neck?*
  ✔ *Did you bend from somewhere in your back?*

ACTION Begin to make this movement again, but this time bend forward using your hip joints. **(6.15)** Keep the vertical alignment of your back, neck and head as you bend forward from your hip joints.

TIP If needed, widen your legs to give your pelvis and upper body the freedom to bend forward.

STOP *Take a break before continuing with Part 2 on the next page.*

figure 6.14

figure 6.15

Bending while sitting: Part 2

👣 Once again, sit vertically balanced.

ACTION Bend forward again and notice what happens to the weight on the bottom of your feet.
- ✔ *Can you feel the weight increasing as you bend forward?*

TIP Your feet play an important role in supporting your weight and movement forward.

ACTION Continue to bend from your hip joints, bringing your nose closer and closer to the imaginary table sitting just out in front of your knees. **(6.16)**
- ✔ *How does your back, neck and head feel bending in this way?*

ACTION Stop bending from your hip joints and bend forward using your back or neck. Compare how bending in this way feels to bending forward using your hip joints.

REST

👣 This time sit facing your therapy table. Sit vertically balanced with your weight on your sit-bones, posterior thighs and feet.

ACTION Reach with your hands toward your table and bend forward using your hip joints. **(6.17)** Keep the alignment of your back, neck and head.

■ Notice the freedom of movement you have in your shoulders, arms and hands when bending from your hip joints.
- ✔ *How does your back feel?*
- ✔ *Your neck?*

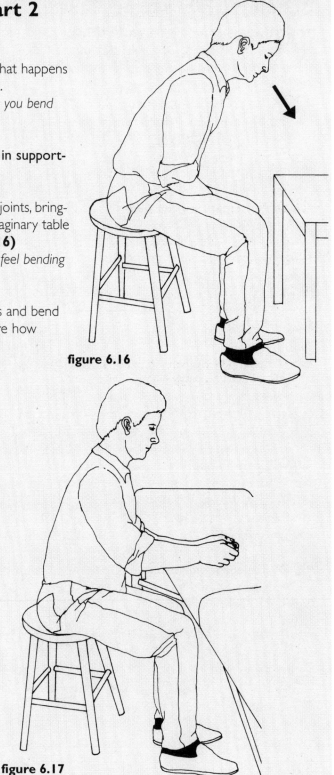

figure 6.16

figure 6.17

ACTION For a moment, reach toward your table, bending forward using your back. **(6.18)**

✔ *Can you feel the lack of movement in your shoulders, arms and hands?*

✔ *Can you sense the increase of muscular tension in your upper body?*

ACTION Go back to bending from your hip joints and begin to reach your hands toward the left side of your table. **(6.19)**

✔ *Can you feel the weight increasing in your left foot?*

ACTION Now reach toward the right side of your table, bending from your hip joints.

✔ *Can you feel the weight increase in your right foot?*

ACTION Begin to alternate reaching toward one side of your table and then toward the other side.

TIP Bending from your hip joints allows you to easily reach in any direction.

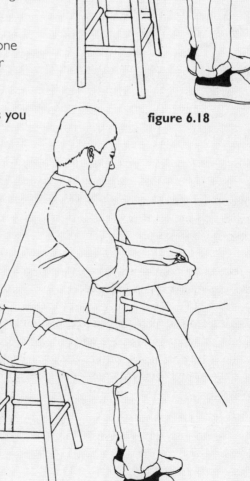

figure 6.18

figure 6.19

STOP *Take a break before continuing with Part 3 on the next page.*

128

TRY THIS

Bending while sitting: Part 3

Sit facing your therapy table. Sit vertically balanced with your weight on your sit-bones, posterior thighs, and feet.

ACTION Imagine you are working with a client, reaching in any direction you would like.

TIP Bend using your hip joints, keep your back, neck and head aligned and let your feet and legs support your weight.

■ Feel the ease of movement you have in your shoulders, arms and hands.

ACTION Stop for a moment and bend without the support of your feet and legs.
   ✔ *Can you feel the muscles of your back and neck working to support your weight?*
   ✔ *Can you feel the effort in your shoulders, arms and hands?*

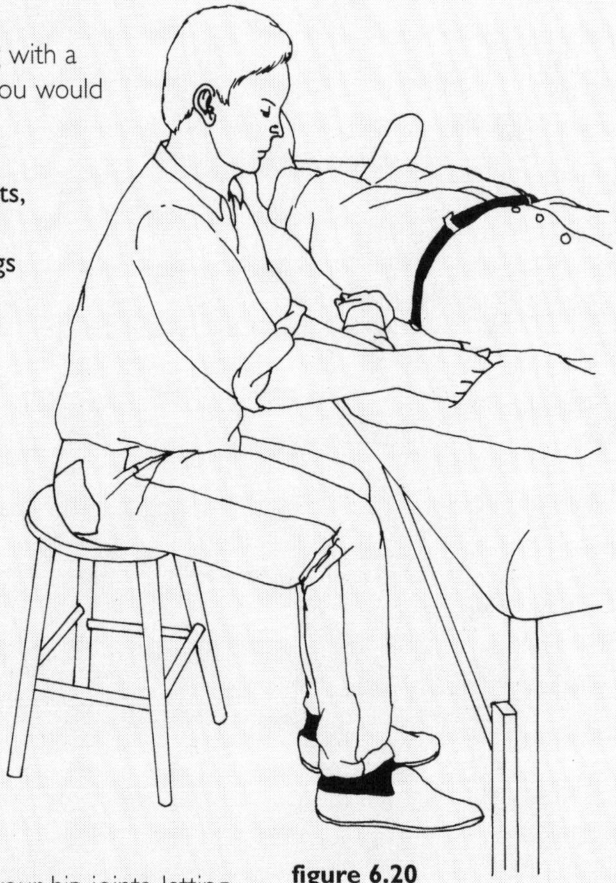

ACTION Once again, bend from your hip joints, letting your legs and feet support you. **(6.20)**

TIP Bend from your hip joints and allow your legs and feet to support you to keep your back, neck and head free from strain and discomfort. This will allow your shoulders, arms and hands to move with ease!

figure 6.20

BENDING SUMMARY

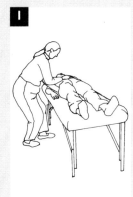

1

Bend from your true hip joints.
This allows you to recruit the strong and powerful muscles of your pelvis and legs.

2

Counterbalance.
Counterbalance your head and pelvis to keep your center of gravity over your legs and feet.

3

Keep your knees in alignment with your feet.
Ideally, your knees should remain over your feet, but when this is not possible, keeping your knees in alignment will help increase your skeletal support and decrease stress in your knee joints.

4

When sitting...
bend from your hip joints and use your legs and feet for support. This will decrease the stress and discomfort in your upper body.

Notes:

PREVENTION TIP

If you are bending throughout most of your sessions, stop and rest occasionally. Resting for a few moments during your sessions gives your body and mind a break. No matter how comfortable and effective your bending body mechanics are, resting once in awhile will keep you energized and relaxed at the same time!

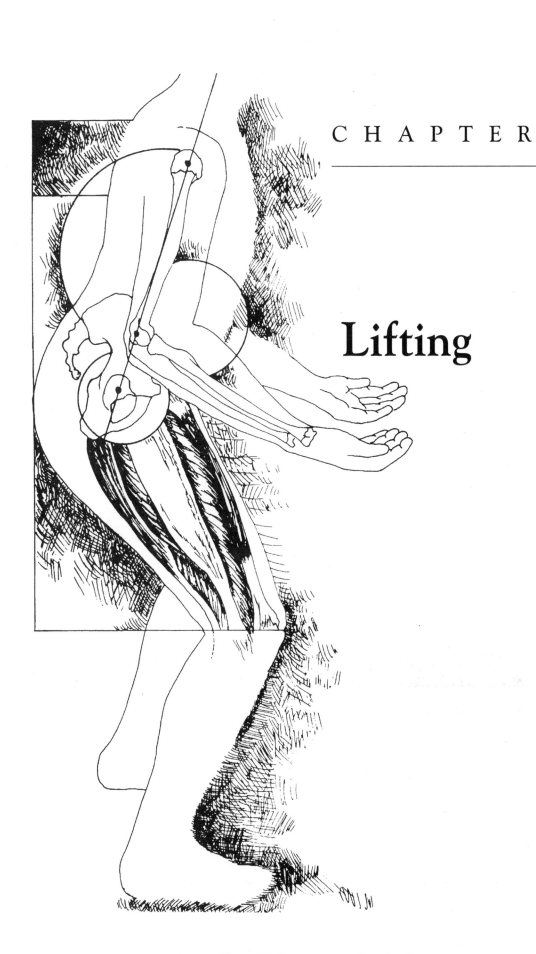

CHAPTER 7

Lifting

INTRODUCTION

Lifting is a common function in manual therapy. Whether you lift frequently throughout your treatments or just occasionally, it is vital that you learn how to lift in a safe and comfortable manner.

Many manual therapists find they feel great about their body mechanics until they attempt to lift. Unfortunately, because of a previous lifting injury or just feeling uncertain, some therapists avoid lifting altogether.

In this chapter you will learn and experience the importance of getting close, using the power of your lower body, and facing the proper direction. You will also learn how to keep your back and upper body stress-free when lifting by lifting, holding and moving a weight using the support of your entire body. Ultimately, you will look forward to lifting with confidence.

Note: Throughout this chapter, the term "weight" will refer to the body parts of your clients, for example a limb or the head.

In this Chapter

THE HABITS OF EVERYDAY LIFE . . .

How do you normally lift throught the day?

◆ Your pets?

◆ A potted plant?
◆ Furniture?

◆ Do you lift with your knees?
◆ Your hip joints?
◆ Your back?

◆ Are you comfortable lifting?
◆ Do you avoid lifting?

AS A MANUAL THERAPIST . . .

How do you lift?

◆ Do you use your back?

◆ Hip joints?

◆ Knees?

◆ Do you lift frequently during your sessions?

◆ Do you avoid lifting?

◆ Do you ask your clients to help you?

◆ Do you lift from a rotated position?

◆ Are you able to hold and move a limb after you have lifted it?

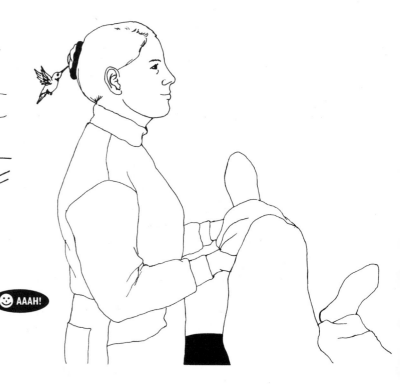

Getting Close

When lifting, it is very important that you get as close to the weight as possible, without losing your own stability. (**7.1a, 1b**) Getting close lessens the amount of work you need to do and decreases your muscular strain and effort.

figure 7.1a

figure 7.1b

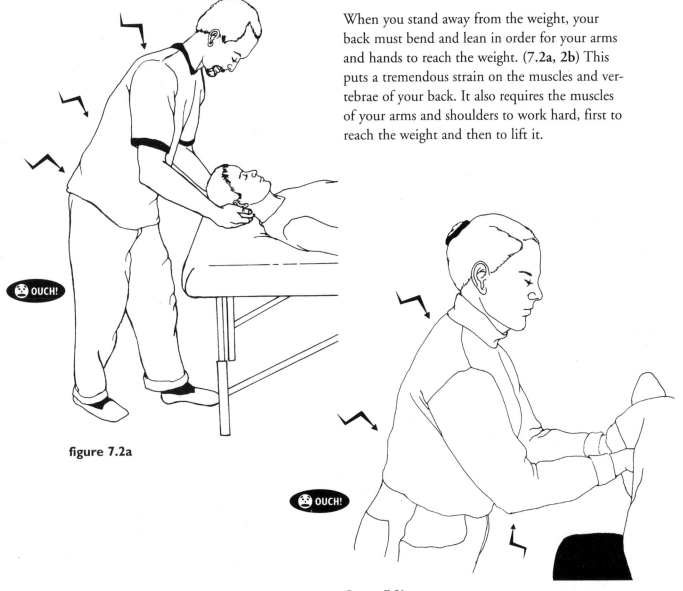

figure 7.2a

figure 7.2b

When you stand away from the weight, your back must bend and lean in order for your arms and hands to reach the weight. (**7.2a, 2b**) This puts a tremendous strain on the muscles and vertebrae of your back. It also requires the muscles of your arms and shoulders to work hard, first to reach the weight and then to lift it.

CHECK THIS OUT

When holding a weight away from your body, your body perceives the weight to be at least ten times heavier then it actually is!

138

Lifting from far and near

 Place a heavy book, for example a telephone book, on the end of your table. Stand above the book. Stand several inches away from your table, but close enough so you can reach the book.

ACTION Lift the book with both hands and hold it up for a few seconds. **(7.3)**

■ Sense the muscular effort you are using.
  ✔ *Can you feel the strain this distance puts on your back, neck and shoulders?*

■ Sense how heavy the book feels to you.
  ✔ *Does it feel heavier then it actually is?*

■ Notice your breathing.
  ✔ *Can you breathe comfortably while lifting from this distance?*

figure 7.3

REST

 This time stand close to your table, above the book. Remain stable on your feet and do not lean into your table.

ACTION Lift the book with both hands. **(7.4)**

■ Sense the amount of muscular effort you are using at this distance.
  ✔ *Has the muscular effort decreased in your back, neck and shoulders?*

■ Sense how heavy the book feels to you now.
  ✔ *Is it lighter or heavier than it was before?*

■ Notice your breathing.
  ✔ *Can you breathe more comfortably?*

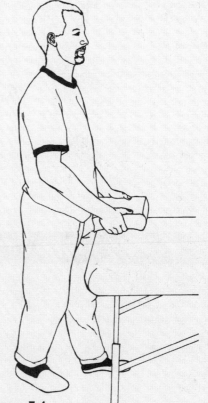

figure 7.4

Let Your Legs Lift It

"Lift with your legs, not your back" is a phrase commonly used when discussing the body mechanics of lifting. It is excellent advice and should be applied to the body mechanics of manual therapists as well.

Lifting with your legs means you are using the power of your lower body to lift the weight. Instead of bending from your back and putting strain on your spine and upper body, bend from your hip joints and knees, and use the power of your legs. This saves your back and upper body from injury and allows the larger and stronger muscles of your lower body to do the work.

Before you begin to lift a weight, bend from your hip joints and your knees. (7.5)

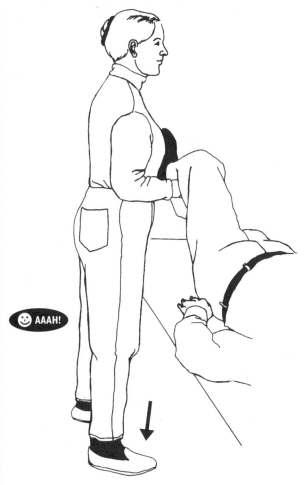

figure 7.6

As you lift the weight, press your feet into the floor and straighten your legs, without locking your knees. (7.6) (Basically, you are raising your body to raise the weight.) Lifting in this manner allows your back to stay upright, and reduces excessive muscular effort in your shoulders, arms and hands allowing them to easily facilitate the lifting.

To lower the weight, bend your knees and return to your original position.

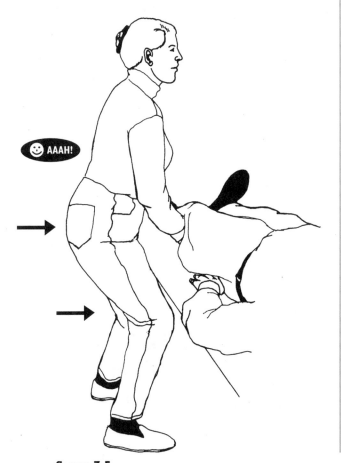

figure 7.5

Face It

Directly facing the weight helps to keep your body in alignment, decreasing your strain and effort. Keep your feet, pelvis and upper body pointed in the direction of the weight to insure you are always facing it. (7.7)

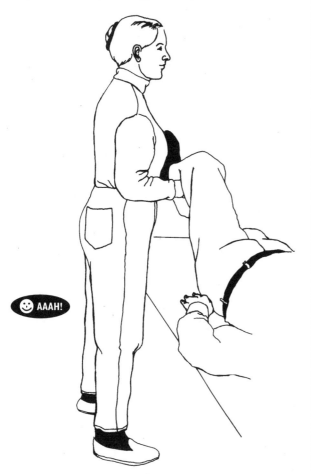

 AAAH!

figure 7.7

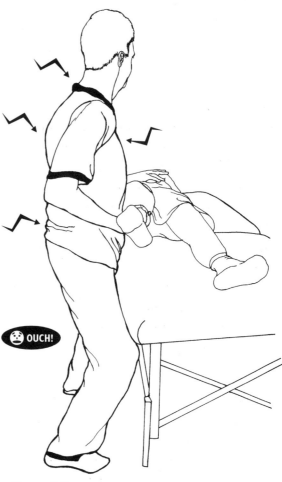

OUCH!

figure 7.8

When your body is twisted while lifting, the vertebrae, disks and soft tissue of your back are under significant abnormal pressure. (7.8) When the weight is light (for example a small arm), lifting from a rotated or twisted stance may seem like the best option; however, lifting any amount of weight from this position puts tremendous strain on your body.

PREVENTION TIP

Many therapists use the short cut of starting their lift by facing the direction of the intended movement. This is where they get into trouble because they are lifting while standing in a rotated or twisted position. It may take a little extra time to initially face the weight, reposition yourself and then move, but it will ultimately save you from pain and injury.

Life After Lifting

Once you have lifted the weight, chances are you will move it or hold it in a position for a period of time. Remain facing the weight if you are going to hold it in one position without moving it in another direction. (7.9) This will keep your body safe from the injuries of twisting and will allow you to hold it as long as you need without becoming uncomfortable. However, never hold a weight longer then you feel is comfortable. Always rest when you need to.

When *holding* the weight, let your upper body simply facilitate the holding. Of course the muscles of your shoulders, arms and hands will be active, but the major work of supporting the weight should be done with your lower body.

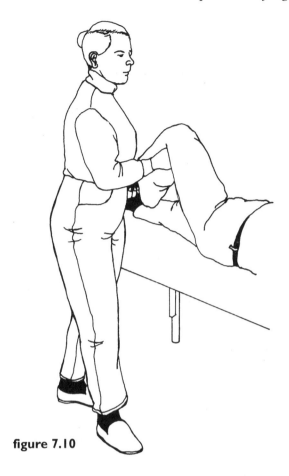

figure 7.10

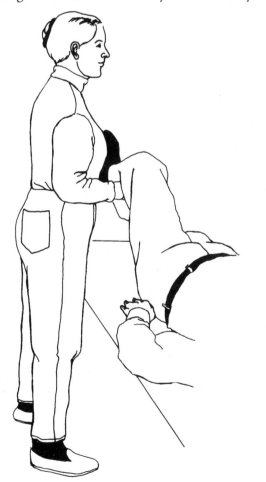

figure 7.9

To *move* the weight in any direction, reposition yourself while holding the weight. (7.10) (If you cannot comfortably reposition yourself while holding the weight, there is a good chance the weight is too heavy for you. If this is the case, ask for help.)

Make sure your entire body is facing the direction of the movement so you are not moving from a rotated stance. Allow the movement of your entire body to facilitate the movement of the weight.

142

Lifting, holding and moving

Have your partner lie supine on your table. Stand beside your table, facing their lower leg. Make sure your entire body, including your feet, is facing their leg.

ACTION Slowly begin to lift their leg. Keep your back upright, bend from your hip joints and your knees. **(7.11)** As you lift their leg, straighten your legs. Lift and lower it a few times.

■ Sense your feet pressing into the floor as you lift.

TIP By pressing your feet into the floor and straightening your legs, you are using the strength and power of your lower body to lift your partner's leg. **(7.12)** This allows your upper body to relax and comfortably facilitate the lifting without strain and effort.

ACTION As you lower their leg, bend your knees.

REST

Ask your partner for feedback.
✔ *How did using your legs affect the lifting of their leg?*

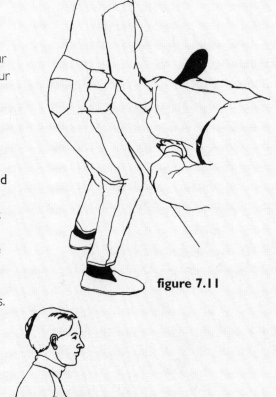

figure 7.11

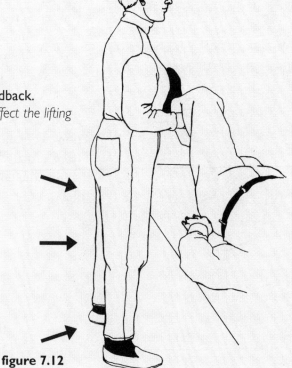

figure 7.12

ACTION Lift your partner's leg again. This time hold it up. In a moment you are going to move their leg and bring their knee toward their chest. Reposition yourself by turning your feet and body so you are facing the intended movement. **(7.13a)**

(Before you start to move be sure that your body is not twisted or rotated.)

ACTION Slowly begin to move your partner's leg and bring their knee toward their chest.

TIP Take as many steps as needed to advance your entire body in the direction of the movement. This will help maintain your vertical alignment, keep you close to the weight and reduce the muscular effort and strain in your upper body. **(7.13b, c)**

ACTION As long as you are moving their knee toward their chest, remain facing the direction of the movement. When you are ready to return the leg to the table, slowly straighten their leg, reposition yourself to face the leg and lower it down.

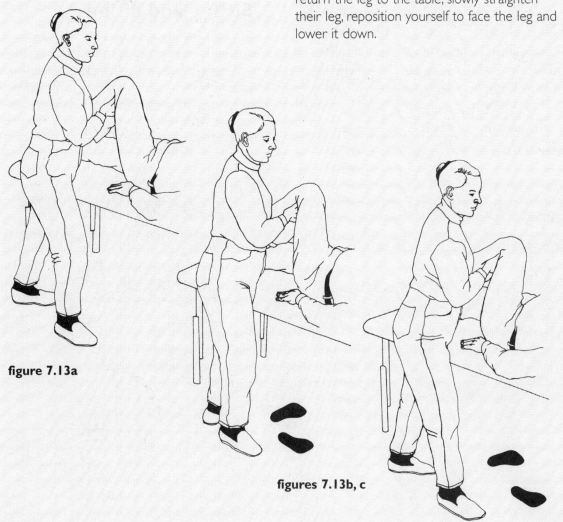

figure 7.13a

figures 7.13b, c

Ask For Help

Before you start to lift any amount of weight, check in with yourself. Make sure you feel comfortable lifting it. Ask your client to help you if you are uncomfortable lifting the weight. You can ask them to move closer to you and to help with the initiation of the lifting. If you feel uncomfortable once you are holding the weight, put it down! Your first priority is your body's comfort and safety.

When utilizing sheets and/or supports, such as bolsters, blocks or rollers, do not hesitate to ask your client to help you. Positioning sheets and supports while lifting can be challenging. Your client can help lift themselves, as long as they are able. Remember, an important part of healthy body mechanics is knowing when to ask for help!

Sitting and lifting

Sit close with your body facing the weight. Press your feet into the floor as you lift up with your hands. This will help you to use your legs for support throughout the lifting. Keep your shoulders, arms and hands relaxed.

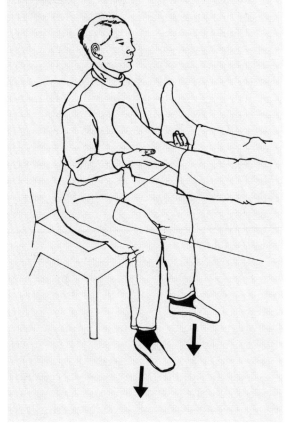

LIFTING SUMMARY

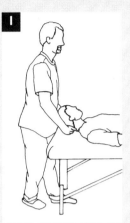

1

Get close.
Stand with your entire body as close to the weight as possible, without leaning and compromising your stability.

2

Lift with you legs, keeping your back upright.
Bend from your hip joints and knees, using the power of your legs, instead of bending from your back and putting strain on your spine and upper body.

3

Initially lift by facing the weight.
Point your feet, pelvis and upper body in the direction of the weight.

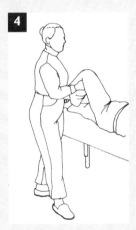

4

Reposition yourself
once you have lifted the weight and are ready to move in a different direction. Make sure your entire body is facing the direction of the movement so you are not moving the weight from a rotated stance.

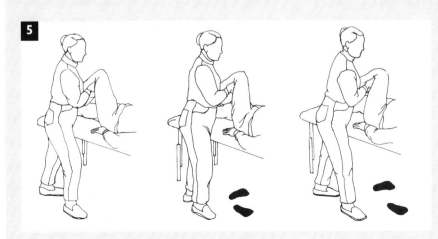

5

Take as many steps as needed
to advance your entire body in the direction of the movement. This will help maintain your vertical alignment, keep you close to the weight and reduce the muscular effort and strain in your upper body.

Notes:

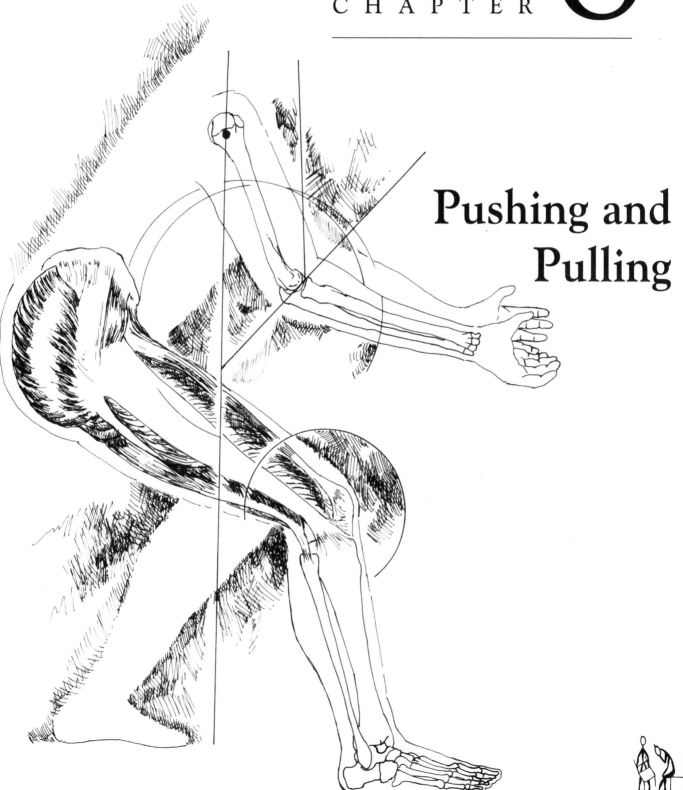

CHAPTER 8

Pushing and Pulling

INTRODUCTION

Pushing and pulling, used together or separately, are major functions of all manual therapies. Whether manipulating bone, soft tissue or energy, the movements of pushing and pulling are inherent in the process. Sometimes the actions of pushing and pulling are obvious, as in chiropractic adjustments, massage therapy strokes and physical therapy manipulations. With other modalities, such as cranial-sacral work, myofascial release, and energy work, pushing and pulling are used in more subtle, but still very critical ways.

In this chapter you will learn how to remain self-supported, gain strength and power from your lower body and maintain your skeletal alignment, no matter what specific type of pushing or pulling you use.

First we will discuss pushing, then pulling, and finally you will have the opportunity to use pushing and pulling together.

THE HABITS OF EVERYDAY LIFE...

How do you normally push throughout the day?

◆ To open or close a door?

◆ A lawn mower?

◆ A shopping cart?

◆ Do you mainly use the strength of your upper body?

◆ Do you primarily use your hands?

◆ Arms?

◆ Shoulders?

◆ Do you consciously involve your lower body?

◆ Are you aware of how often you push?

AS A MANUAL THERAPIST...

How do you push?

◆ Do you push with constant or thrust force?

◆ Do you use your arms and hands?
◆ Do you use your fingers?
◆ How do you use your upper and lower body?

◆ Do you push more than you pull?

◆ Do your hands and fingers get sore or stiff when you push?

◆ How does pushing affect your breathing?

◆ Can you maintain your balance throughout the act of pushing?
◆ Can you maintain constant pressure and contact through the full act of pushing?

Self-Supported Pushing

When pushing, stand self-supported on your feet, without relying on the body of your client to hold you up. When you are self-supported, your feet support the weight of your body, your legs form a strong angle of support, and your upper body is free to move. (8.1)

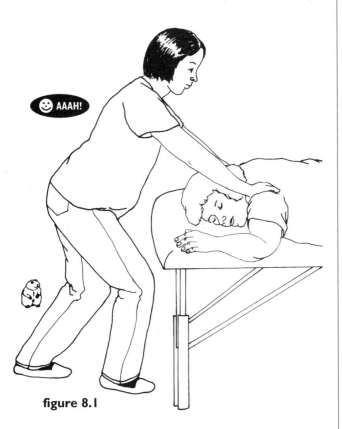

figure 8.1

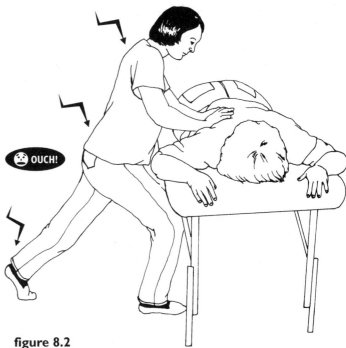

figure 8.2

When you lose or compromise your stability and support, you decrease the effectiveness of your pushing. Relying on the body of your client for support, rather than using your body to support you, compromises your stability and support. (8.2) Your hands are holding you up primarily by way of your client's body. This causes tremendous strain to your hands and wrists, putting them at risk for injury. The muscles of your lower and upper body must work hard to adjust for your lack of balance and self support. This also puts your client at risk for injury, because your hands do not have the control needed for sensitive touch, and your body cannot quickly respond to change direction or pressure.

PREVENTION TIP

Support yourself with your feet to help prevent injury and pain in your hands and wrists. This will decrease the muscular effort in your upper body, such as your back and shoulders, and helps to ensure that your touch is always appropriate, e.g., not too deep.

PARTNER PRACTICE

Self-supported pushing: Part 1

 Have your partner lie prone on your therapy table. Stand at the end of your table, above their head.

ACTION Push into your partner's back, standing in a way that requires you to use their body for your support. **(8.3)**
   ✔ *Sense how your hands and wrists feel pushing in this way.*
   ✔ *How much control and sensitivity do you have in your hands right now?*

■ Notice the effort in your back and shoulders.
   ✔ *Is there an increase of muscular effort in your back?*
   ✔ *In your shoulders?*

■ Notice the effort in your legs and feet.
   ✔ *Are your feet able to remain in contact with the ground?*
   ✔ *Do you feel balanced and in control?*

■ Notice if you are able to breathe comfortably.

ACTION Now begin to pull away from your partner's back.

figure 8.3

■ Sense how much effort it takes to pull away from your pushing.
   ✔ *Where is this effort centered?*
   ✔ *Your back?*
   ✔ *Your legs?*

 Ask your partner for feedback.
   ✔ *How did it feel to have you supporting yourself with their body?*
   ✔ *How much control and sensitivity could they feel in your hands?*

STOP *Take a break before continuing with Part 2 on the next page.*

PARTNER PRACTICE

Self-supported pushing: Part 2

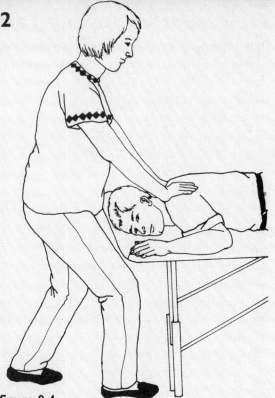

Stand in a self-supported way, with your weight equally distributed between both feet. (Standing, page 72) **(8.4)**

ACTION Push into your partner's back.

■ Sense how your hands and wrists feel.
  ✔ *Has the control and sensitivity in your hands increased?*

■ Notice your back and shoulders.
  ✔ *Has the effort decreased?*

■ Notice your legs and feet.
  ✔ *Are your feet able to remain in contact with the ground?*
  ✔ *Do you feel balanced and in control?*

ACTION Now, slowly begin to push your hands down your partner's upper back. **(8.5)**

figure 8.4

Remember to bend from your hip joints and knees, letting your back remain in a neutral position. (Bending, page 119)

■ Sense the amount of control and balance you have while pushing your hands down your partner's back.

■ Sense the amount of sensitivity you continue to have in your hands.

REST

Ask your partner for feedback.

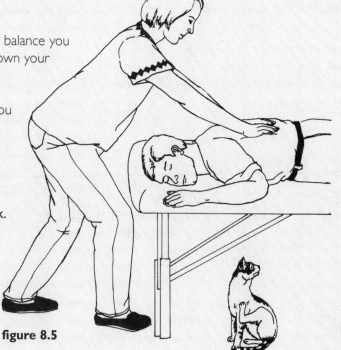

  ✔ *How did standing in a self-supported way affect your pushing?*
  ✔ *Could they sense the increase of control and sensitivity in your hands?*

figure 8.5

Stand in a way that requires you to use their body for your support.

ACTION Begin to push your hands down their back.

■ Notice how much effort it takes to push your hands down your partner's back while standing in a non-supported way.
✔ *What would happen if you were to quickly remove your hands from your partner's back? (You would probably fall into them, right?)*

REST

Ask your partner for feedback.
✔ *How did pushing your hands down their back feel this time?*

ACTION Once again stand self-supported and push your hands down your partner's back.

ACTION Quickly remove your hands from your partner's back **(8.6)**
✔ *Are you able to remain on your feet without falling onto your partner?*

Congratulations! You are pushing in a self-supported manner and increasing your overall control and sensitivity.

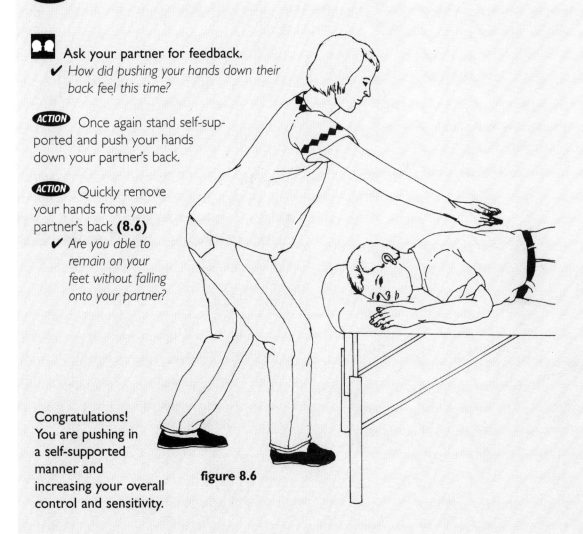

figure 8.6

The Pushing Power of Your Lower Body

Pushing is a physically demanding part of your body mechanics, especially when you need to push with strength, e.g., to apply depth or pressure. When you use your *lower body* to generate the strength you need, you keep your upper body stress free.

You have learned that your hip joints are the strongest and most stable joints in your body. These strong joints are critical when it comes to pushing. Bend from your hip joints while pushing to align your legs and feet in a very powerful way. This allows your feet to press into the floor as you push your upper body forward, increasing your power and strength tenfold. (**8.7**) (This feels very different compared to leaning your weight into your client to increase your power.)

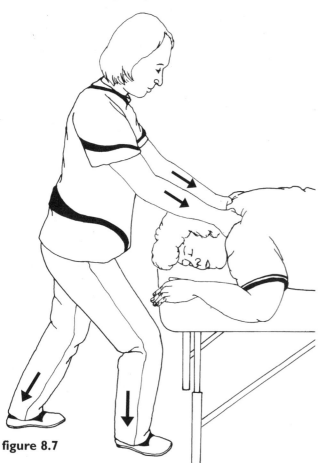

figure 8.7

CHECK THIS OUT

A track runner starts a race by bending from her hips joints, bending her knees and placing her feet into the starting blocks. This allows her to press her feet into the blocks in an explosive way in order to propel her body forward to begin running.

PREVENTION TIP

Use your lower body to generate power and decrease the strain and stress on your upper body. If your upper body is primarily used to generate the power, the muscles and joints of your shoulders, arms and hands will quickly fatigue and become strained.

Using the larger muscles and joints of your lower body for power saves the smaller muscles and joints of your upper body from injury!

PARTNER PRACTICE

Powerful pushing: Part 1

 Have your partner lie prone on your table.
Stand above their head.

■ Think about how you will push into your partner's back using your feet, legs and pelvis to generate the amount of strength and depth you desire while you remain self-supported.

ACTION Bend from your hip joints, bend your knees and let both of your feet press into the floor as you slowly begin to push into your partner's back. **(8.8)** Do not move down their back for now, just push into their back, hold for a moment and then release the pressure.

ACTION Practice this several times.

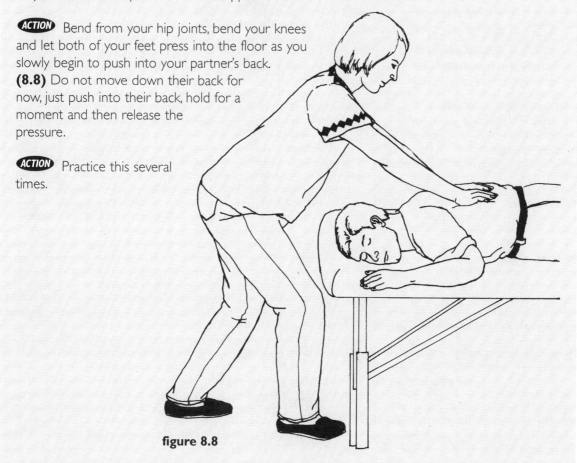

figure 8.8

STOP *Take a break before continuing with Part 2 on the next page.*

Powerful pushing: Part 2

Stand again above your partner's head.

ACTION Push into their back. To increase the power of push, increase the amount of press into the floor with your feet. **(8.9)**

TIP It is important that both of your feet are pressing equally. Notice how your feet respond to pressing into the floor. If, for example, you find that your back foot tends to raise up or has less power, increase the bend in your hip joints.

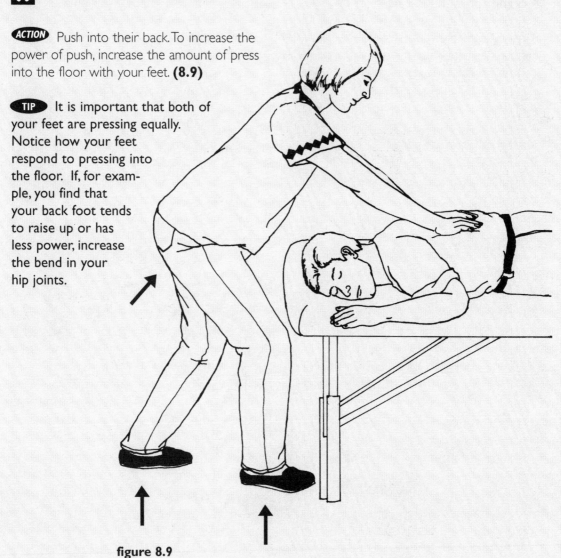

figure 8.9

ACTION Move your hands down your partner's back while you push. **(8.10)** Continue to use your feet and lower body to generate the strength you need.

ACTION As you move down their back and elongate your body, maintain equal foot pressure—do not move so far that you are no longer self-supported.

■ Focus on incorporating the movement of your entire body into your pushing.

TIP When you push and generate the power you need with your feet, your shoulders, arms and hands can relax and simply facilitate the pushing.

Ask your partner for feedback.
✔ *Did pushing in a self-supported way and pressing your feet into the floor increase the effectiveness of your pushing?*

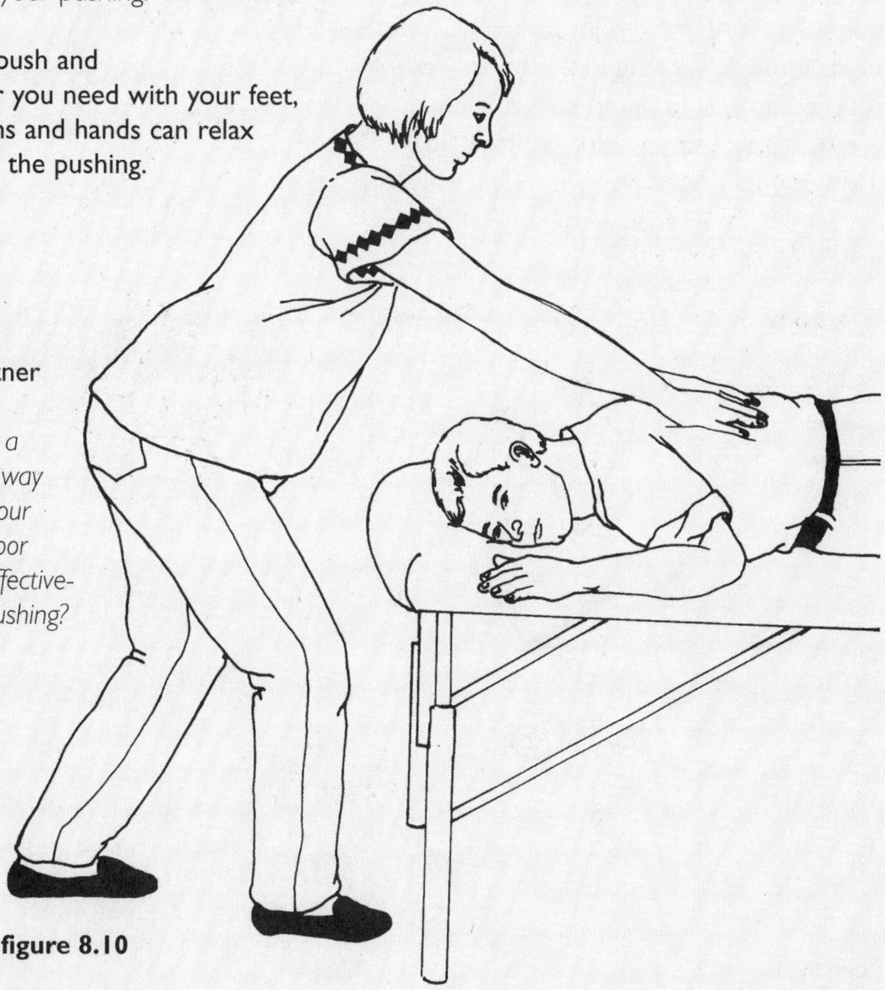

figure 8.10

Pushing With Alignment

Though all of the points we have mentioned so far are important, if you don't maintain your skeletal alignment, you may still lack the effectiveness and comfort you desire.

When pushing, you are generating and transmitting force from your feet up to your hands. If the force you create travels up through a well aligned skeleton, your joints will remain healthy. (**8.11**) However, if the force you create travels through a misaligned skeleton, stress will centralize at the point of misalignment. Consequently, your muscles, tendons, ligaments and the rest of your soft tissue will work very hard to accommodate the lack of alignment. (**8.12**)

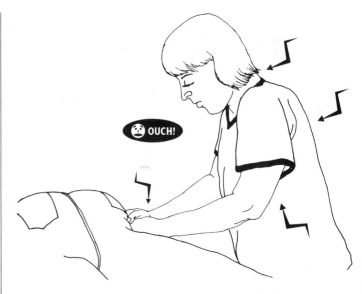

figure 8.12

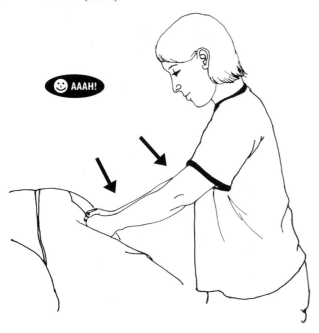

figure 8.11

CHECK THIS OUT

Babies learns how to push by using their skeletal alignment. Lying on her stomach, at the age of about 6 months, a baby will push her upper body up with her hands. By doing this, the baby can see around her environment more easily.

The fact that a baby can push herself up in this way is amazing considering she has very little muscle tone. What she does have is skeletal strength which she intuitively knows how to use. The baby positions herself in such a way that she uses the strength of her aligned skeleton to do the pushing—she creates a "bridge" of support with her hands, arms and chest.

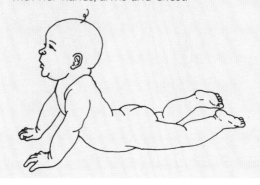

TRY THIS

Pushing up

Lie on your stomach on the floor. Place your hands in a "push up" position, shoulder width apart.

ACTION Slowly begin to push your upper body away from the floor by pushing your hands into the floor. **(8.13)**

TIP Make sure your hands are in alignment with your elbows and your elbows are in alignment with your shoulders. In this position you create a "bridge" of support.
- ✔ *Can you sense the strength of your skeletal alignment as you push yourself away from the floor?*
- ✔ *How much muscular effort is needed?*

ACTION Move your hands outside the width of your shoulders.
- ✔ *Does this make the pushing easier or harder?*
- ✔ *How much muscular effort is needed?*

ACTION Move your hands inside the width of your shoulders.
- ✔ *Does this make the pushing easier or harder?*
- ✔ *How much muscular effort is needed?*

ACTION Bring your hands back underneath your shoulders and push yourself up again.

TIP In this position you are using your skeletal alignment to your advantage, just has you did when you were a baby! **(8.14)**

figure 8.13

figure 8.14

As you experienced in the last lesson, pushing yourself away from the floor can help you sense the importance of skeletal alignment. When it comes to using manual pushing techniques, you can use the same principle. When you find the strongest angle of alignment for yourself, your muscular effort will decrease and you will use the strength of your skeletal alignment. (8.15 a,b)

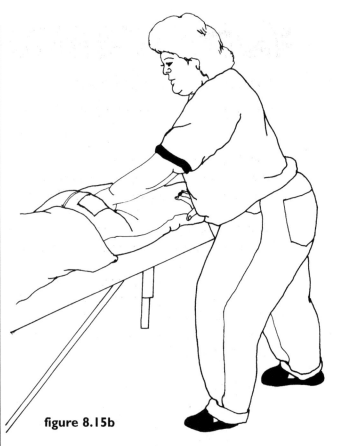

figure 8.15b

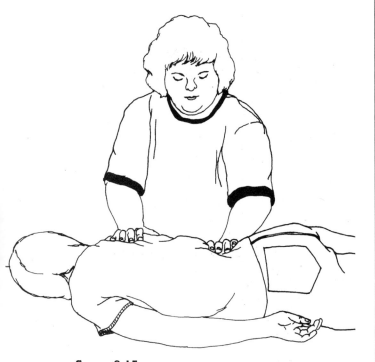

figure 8.15a

Pushing a book: Part 1

Place a heavy book, for example a telephone book, on one end of your therapy table.

 Stand behind the book and place your hands on it as if you were going to push the book away from you, down toward the other end of the table. **(8.16)**

ACTION Push the book down the table. Each time the book moves beyond your reach, bring it back and start again.
   ✔ From where in your body are you pushing?
   ✔ Are you primarily pushing with your hands and wrists?
   ✔ Are you pushing from somewhere in your back or shoulders?

■ Notice from where you are bending.
   ✔ Are you bending from your back?
   ✔ Are you bending from your hip joints?
   (Bending, page 119)

ACTION Continue to push the book down the table and freeze your position once you have pushed it almost past your reach.

■ Notice how you are supporting yourself.
   ✔ Are you leaning into your hands for support?

ACTION Continue to hold your position and become more aware of your feet, legs and pelvis.

TIP Here is an opportunity to become self-supported. This means your hands remain in solid contact with the book, but instead of supporting yourself through your hands and wrists, you support yourself with your lower body. **(8.17)**

STOP *Take a break before continuing with Part 2 on the next page.*

figure 8.16

figure 8.17

TRY THIS

Pushing a book: Part 2

Stand behind the book and allow your feet and legs to support your body weight.

ACTION Push the book down your table.
✔ *Can you quickly remove your hands and remain stable on your feet?* **(8.18)**

ACTION Begin to use your feet to increase the strength of your pushing. Press your feet into the floor as you push your hands forward.

TIP Here is a great chance to let your lower body "drive" the pushing and allow your upper body to follow the movement. Allow your hands to facilitate the pushing, but let your feet generate the strength behind it.

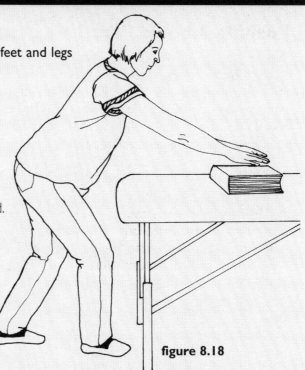

figure 8.18

■ Bring your attention to your shoulders.
✔ *Are you raising your shoulders as you push?*

ACTION Try raising your shoulders so high they are practically underneath your ears. Continue to push the book with your shoulders raised high. **(8.19)**

■ Notice the muscular effort needed when your shoulders are up.
✔ *Can you sense how your head, back, arms, hands and wrists are affected with your shoulders in this high position?*

REST

figure 8.19

ACTION Wiggle your shoulders around a little bit and find a comfortable place for them to rest.

ACTION Push the book down the table again. Allow your lower body to support you and to "drive" the pushing. **(8.20)** Let your head, back, and shoulders relax and go along for the ride.

ACTION Finally, push the book and let your hands and wrists simply facilitate the pushing.

TIP Incorporate the concepts of this lesson into your body mechanics to increase the effectiveness and comfort of your pushing.

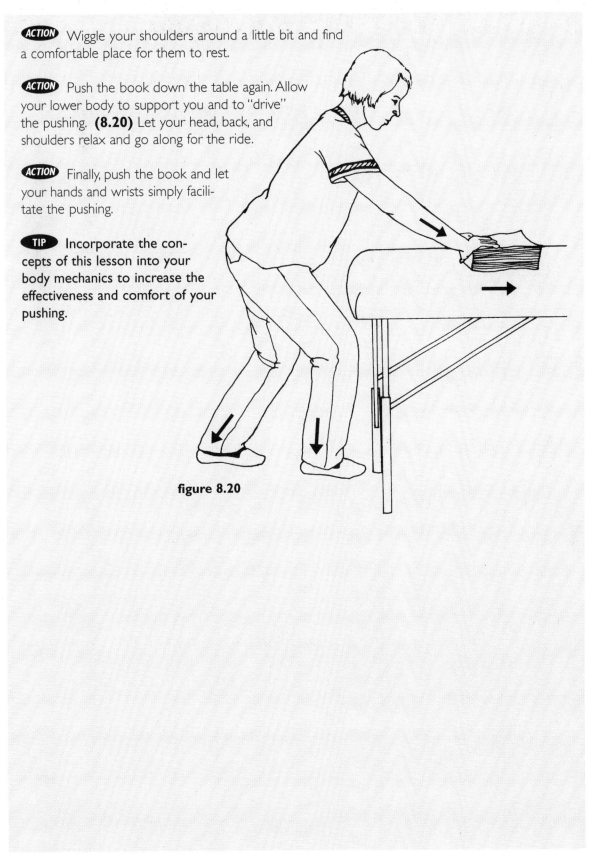

figure 8.20

THE HABITS OF EVERYDAY LIFE...

How do you normally pull throughout the day?

◆ Open a door?

◆ A drawer?

◆ A broom?

◆ A vacuum cleaner?

◆ A rake?

◆ Do you primarily use the strength of your arms?

◆ Your hands?

◆ Your shoulders?

◆ Do you use the strength of your back?

◆ Your neck?

AS A MANUAL THERAPIST...

How do you pull?

◆ Do you use the strength of your hands?

◆ Your arms?

◆ Your shoulders?

◆ Do you use your fingers?

◆ Do your hands and fingers get sore or stiff when you pull?

◆ How does pulling affect your breathing?

◆ Can you maintain your balance throughout the act of pulling?

◆ Do you pull more than you push?

Pulling

Pulling is basically pushing in reverse. You might have heard the phrase "push, don't pull." Ordinarily it is easier to push your body weight forward instead of pulling your body weight backward. However, if you keep in mind the same body mechanics you learned for pushing, along with a few new concepts, you will pull with the same effectiveness and comfort.

Self-Supported Pulling

Remaining balanced and stable is as important while pulling as it is while pushing. However, the tendency while pulling is to lean back and suspend all of the body's weight through the hands, arms and shoulders. (8.21a. b)

figure 8.21b

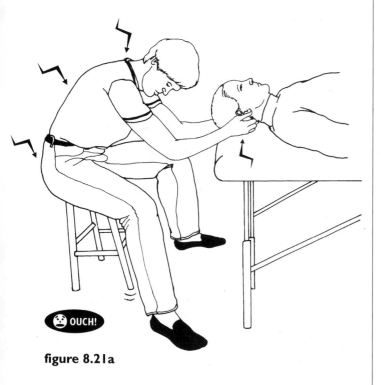

figure 8.21a

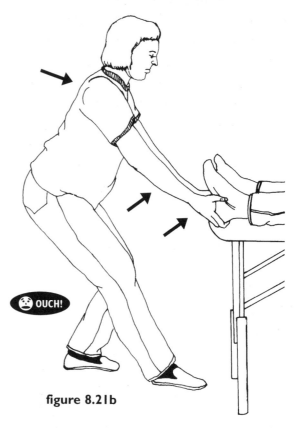

PREVENTION TIP

Pulling only with your upper body requires your hands to grip tightly. This can lead to discomfort and overuse of your hands. Use your lower body for support and allow your hands to pull with sensitivity. This reduces their chance of injury.

Pulling requires the same skeletal support as pushing. Without the support of your feet and legs, the muscles and joints of your upper body must work overtime to support you and pull.

Whether you are pulling subtly with you fingers or pulling forcefully with your hands, you must pull using the support of your entire body to avoid injury.

As you experienced with pushing, integrating your entire body into pulling will increase your effectiveness and keep your body stress-free. (8.22a, b)

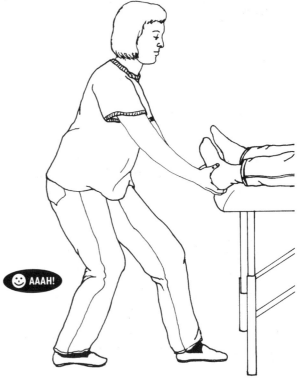

figure 8.22b

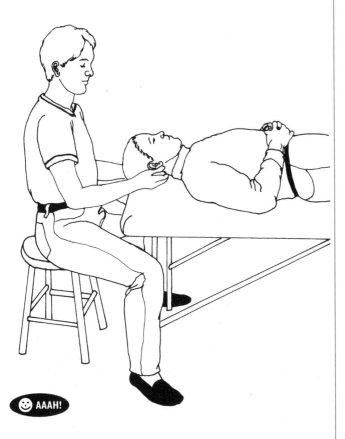

figure 8.22a

CHECK THIS OUT

The function of pulling starts at a very early age. Shortly after birth, a baby will begin to grasp and pull at his mothers breast while feeding. At the age of about 6 months a baby will lie on his back and pull his feet towards his mouth. When's the last time you tried that?

TRY THIS

Self-supported pulling

Stand behind your table. Your feet should be far enough away from your table so your body leans back a bit. **(8.23)**

ACTION Pull your table with your hands. (Ask your partner to hold the other end of your table so it remains still.)

✔ *Can you feel how your body weight is being suspended from your hands?*

TIP In this position, you are relying on the stability of your table and the strength of your hands to keep yourself from falling backward.

✔ *What would happen if you were to suddenly let go of your table? You would probably lose your balance and fall backward!*

■ Notice how pulling in this way increases the muscular effort in your hands, arms and shoulders.

✔ *Does it increase in your neck and back?*

✔ *In your legs?*

figure 8.23

REST

♊ This time, place your feet, legs and pelvis under you in such a way that they support your body weight.

ACTION Pull on your table again, bending from your hip joints and knees, letting your back remain in a neutral position. **(8.24)**

■ Sense the amount of control and sensitivity you have in your hands.
  ✔ *Do you feel less strain on your hands and wrists?*
  ✔ *Can you let your hands pull without overly gripping your table?*
  ✔ *Has the effort in your shoulders decreased?*
  ✔ *Has the effort in your back and legs decreased?*
  ✔ *Are your feet able to remain in contact with the ground?*

ACTION As you pull, focus on remaining stable and balanced on your feet. Quickly remove your hands from your table.
  ✔ *Are you able to remain on your feet without falling backward?* **(8.25)**

TIP Stand self-supported to allow your hands to pull with sensitivity and the large muscles of your lower body to support you.

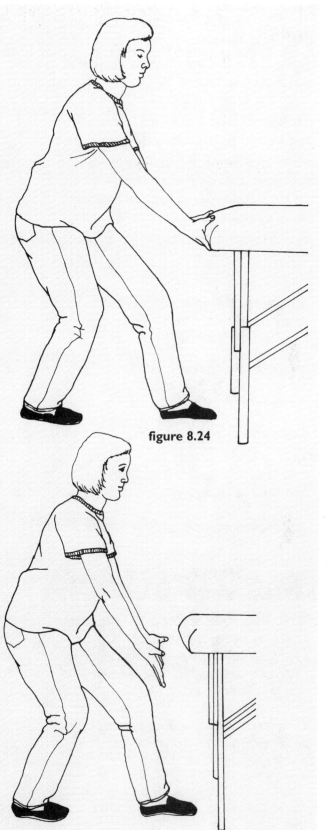

figure 8.24

figure 8.25

Pressing Down and Pulling Back

Have you ever watched the movements of a good rower? As they pull back on the oars with their upper body, they press with their feet, into the boat. The simultaneous pressing with their feet and pulling with their upper body, generates the power they need to move the boat through the water. If you have ever watched a regatta (boat race), you have seen how much speed their pressing and pulling power can generate. **(8.26)**

figure 8.27

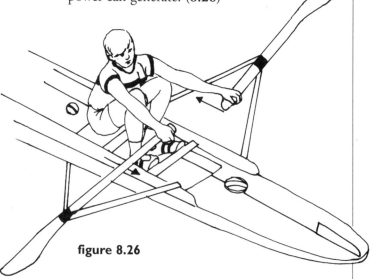

figure 8.26

PREVENTION TIP

Use your entire body for pulling. No matter if you are pulling a large limb or pulling subtly on soft tissue, remain self-supported and generate the power you need with your lower body.

This saves your upper body from overuse and injury and allows your hands to pull with sensitivity.

You can use the same kind of pressing movements to generate power when you pull. With your feet, legs and pelvis supporting you, you can press into the ground with your feet as you pull back with your hands. Pulling in this fashion is a dynamic way to use your entire body. You are not isolating the pulling action in one area of your body, e.g., your hands and arms; rather, you are using your entire body to pull with power and stability. (8.27)

TRY THIS

Press and pull

Stand behind your table in a self-supported manner.

ACTION Bend from your hip joints, bend your knees and let your feet press into the floor as you begin to pull your table. (Ask your partner to hold the other end.) **(8.28)**

TIP If you have difficulty pressing your feet into the floor, try lowering your pelvis down—make the initial movement of sitting down in a chair.

ACTION Let your hands pull with sensitivity—do not overly grip with your hands.
- ✔ *Can you sense how your entire body can be involved with the pulling?*
- ✔ *How much do you need to press your feet into the floor in order to increase your pulling power?*

TIP Coordinate the pressing of your feet with the pulling of your hands. This will allow you to make subtle changes in your pulling.

TIP If you want to pull with power, press your feet into the floor with force. If on the other hand, you want very little power, then you will want to maintain clear contact with the floor, but you will not need to press with much force.

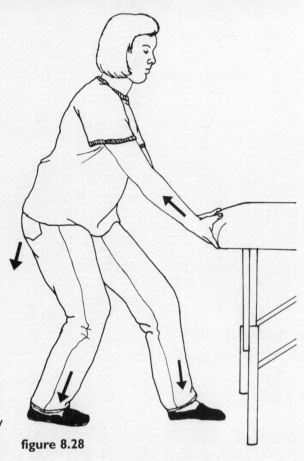

figure 8.28

Pulling a leg

Have your partner lie supine on your table. Stand at the end of your table and place your hands on their foot as if you were going to traction or pull their leg. **(8.29)**

ACTION Begin to pull your partner's leg. (Only pull for a few seconds at a time.)

■ Notice from where in your body you are pulling.
   ✔ *Are you primarily pulling using the strength of your hands and wrists?*

TIP Try to reduce the amount of effort you are using in your hands. If your lower body is supporting you, your hands can facilitate the pulling without overly gripping.

   ✔ *Are you pulling from somewhere in your back or shoulders?*

TIP Let you back and shoulders relax. There is no need for them to work hard if you are supporting yourself with your lower body.

ACTION Continue to pull your partner's leg and freeze your position for a moment.

■ Notice how you are supporting yourself.
   ✔ *Are you supporting yourself without using your partner?*

REST

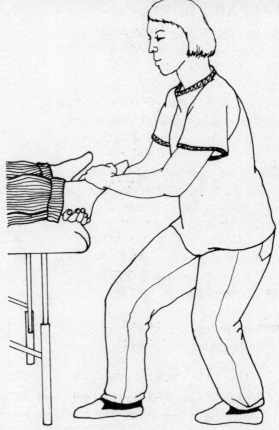

figure 8.29

Stand again at the end of your table.

ACTION Continue to pull your partner's leg and focus on becoming self-supported. Allow your feet and legs to support your body weight. (Standing, page 72)
✔ *Could you quickly remove your hands and remain stable on your feet?*

ACTION Continue to pull, and begin to press your feet into the floor to increase the strength of your pulling. **(8.30)**

TIP While pressing your feet into the floor and pulling, you may feel as if you would like to take a step back. This is fine—just make sure you remain stable on your feet.

ACTION Now decrease the amount of pull by pressing less into the floor.

TIP Even though you are pressing less with your feet and decreasing your pull, continue to remain self-supported.

Ask your partner for feedback.
✔ *How did standing in a self-supported way and using your feet to press into the floor affect the pulling of their leg?*

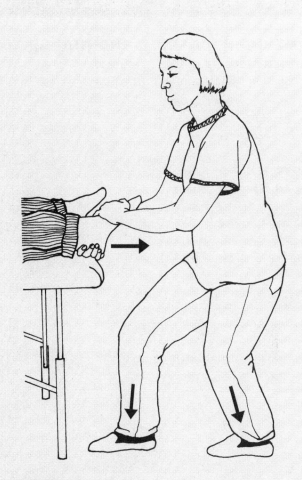

figure 8.30

PARTNER PRACTICE

Bringing pushing and pulling together

During treatments you will use the functions of pushing and pulling together. Now that you have experienced the two separately, you have a clearer idea of how to use your body mechanics to integrate them.

Have your partner lie down on your table.

ACTION Take a few moments and practice using pushing and pulling together.

For example:

- Pulling/tractioning an arm while pushing/stretching a hip. **(8.31)**

- Stretching/pushing and pulling soft tissue. **(8.32)**

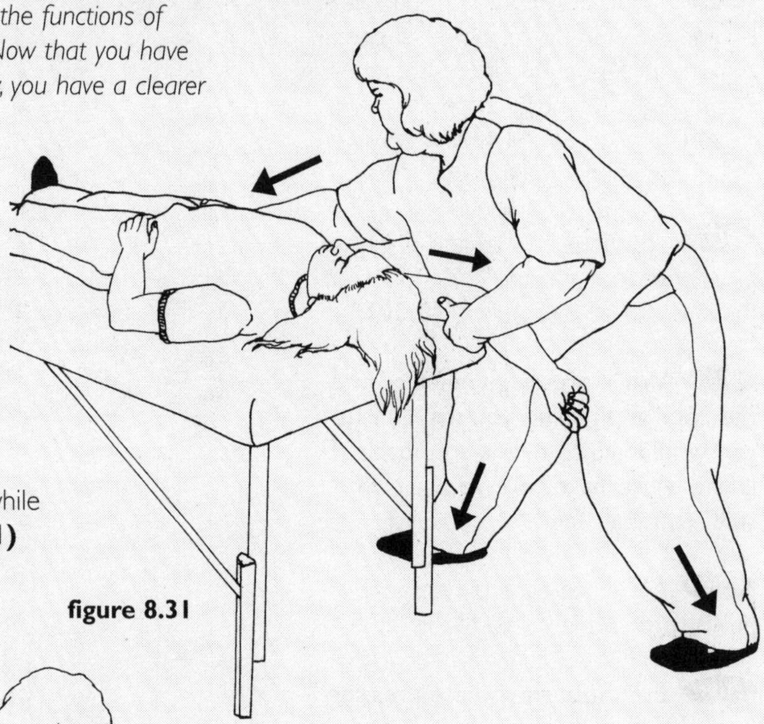

figure 8.31

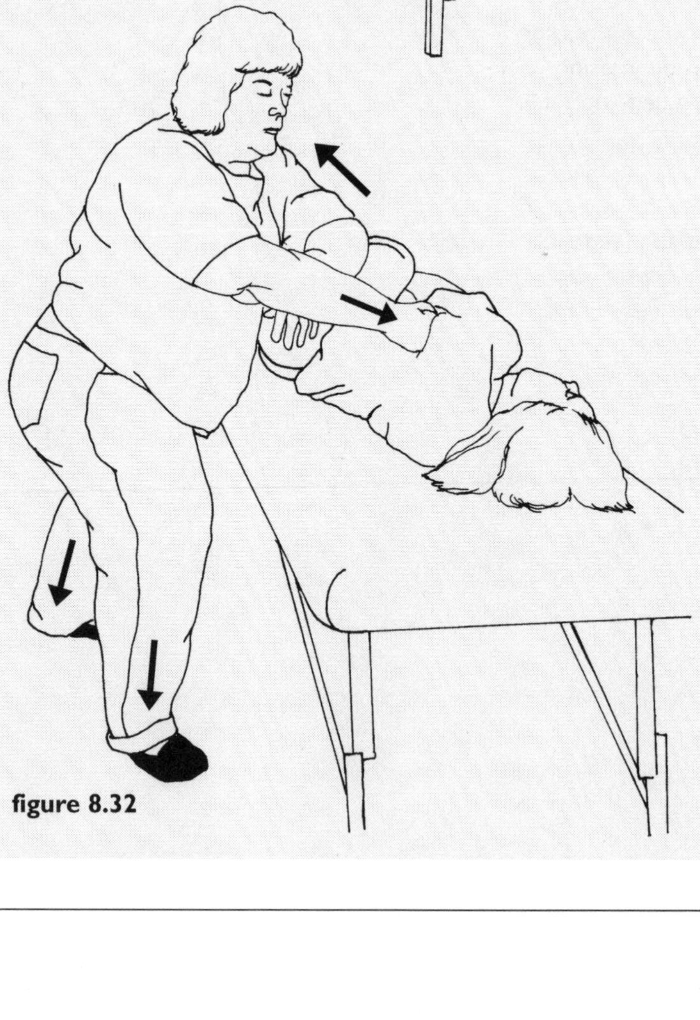

figure 8.32

PUSHING AND PULLING SUMMARY

1

Be self-supported.

When you are self-supported, your feet support the weight of your body, your legs form a strong angle of support, and your upper body is free to move.

2

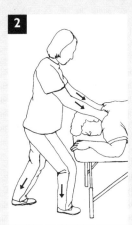

Use the power of your lower body.

Bend from your hip joints while pushing to align your legs and feet in a very powerful way. This allows your feet to press into the floor as you push your upper body forward, increasing your power and strength tenfold.

3

Use the strongest angle of alignment.

When you find the strongest angle of alignment for yourself, your muscular effort will decrease and you will use the strength of your skeletal alignment.

4

Press with your feet as you pull with your hands.

Bend from your hip joints and your knees, letting your feet press into the floor as you pull.

5

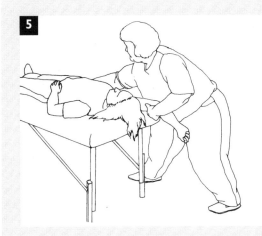

Pushing and pulling together.

Continue to bend from your hip joints and press with your feet as you push and pull.

Notes:

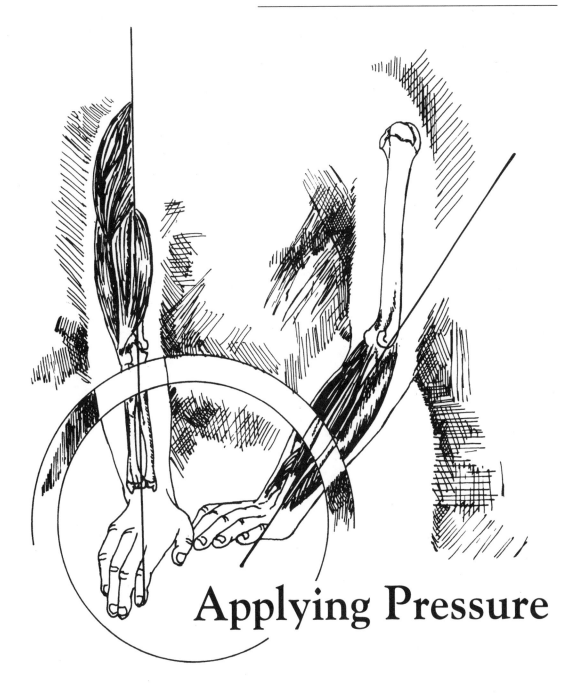

CHAPTER 9

Applying Pressure

INTRODUCTION

No matter what kind of manual therapy you practice, the one function that remains a constant is the application of pressure. Whether you are lightly placing or deeply pressing your hands, forearms or elbows, you are in some degree or another applying pressure. It is often by the manipulations of pressure through which clients are led back to health. Unfortunately for the manual therapist, applying pressure is what often leads to pain and injury.

In this chapter we will focus on applying pressure by using your body weight, finding the most effective angle of alignment and the importance of getting close to your work. We will also look at how your beliefs can influence how you use your body and how breathing can keep your body from "getting stuck" in "static" pressure. All of these important concepts will help increase your effectiveness in applying pressure and help keep your body safe from pain and injury!

THE HABITS OF EVERYDAY LIFE...

How do you apply pressure throughout the day?

◆ To groom your pet?

◆ Waxing your car?

◆ Polishing furniture?

◆ Which hand do you primarily use?

◆ Do you use your fingers?

◆ The palm of your hand?

◆ The heel of your hand?

◆ A combination?

◆ Do you ever use your forearms or elbows?

◆ Do you tend to press hard when applying pressure?

◆ Are you aware of your breathing?

AS A MANUAL THERAPIST...

How do you apply pressure?

◆ Do you primarily use the heel of your hands?

◆ Your fingers?

◆ Your forearms?

◆ Do you mainly use your right hand?

◆ Your left?

◆ Do you use your elbows?

◆ What happens to your breathing when you apply pressure?

◆ Do you find yourself becoming fatigued during or after applying pressure?

Angles of Alignment

Throughout this book, you have learned the importance of maintaining your skeletal alignment. To be honest, if you always kept your skeleton perfectly aligned, you would move stiffly, like some kind of robot. The purpose of good alignment is not to make you into a stiff robot. Rather, the purpose of good alignment is to learn how to use the integrity of your skeletal system to do the work it is designed to do.

An effective angle of alignment is one in which you can transfer your body weight to your shoulders, to your elbows, to your wrist, ultimately feeling the support of your joints. If you feel your muscles working hard to keep your joints stable, then you need to adjust your alignment until you feel your muscular effort has decreased and your skeletal strength has increased. (9.1, 9.2)

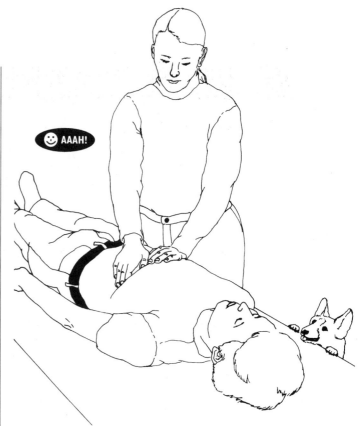

figure 9.1

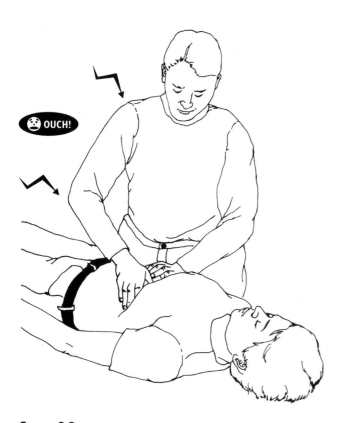

figure 9.2

The bones and connective tissue that make up your skeletal system—even the small bones of your hands—are very strong and stable. When you use these strong bones together, in alignment, they are even stronger. Keep this thought in mind as you apply pressure—when I use my skeleton in alignment, I decrease my chances of injury.

PREVENTION TIP

Applying pressure can cause injury when the joints of the hand, arm and shoulder are poorly aligned. Over time these joints become weak and unable to withstand such force. Keep your joints in alignment or "stacked" to ensure the integrity of your joints and allow the pressure to travel in a direct line.

PARTNER PRACTICE

Finding an effective angle of alignment: Part 1

👣 Ask your partner to lie prone on your table and stand by their side. Stand in a self-supported and comfortably aligned manner. (Standing, page 76)

ACTION Apply pressure, with the top of your fist, to the side of your partner's back. **(9.3)** Find an effective angle of alignment which allows you to press comfortably and effectively from your standing position.

TIP Remember, the most effective angle of alignment is one that keeps your joints supported, as much as possible. It is also an angle in which you can sense the strength of your alignment as you apply pressure.

ACTION Press, using your body weight, through your shoulder into your straight arm and into your fist. (Do not let your wrist flex and be sure to remain self-supported.)

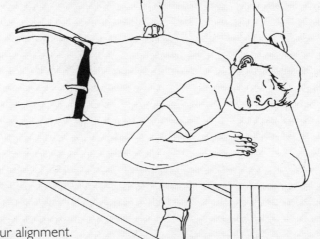

■ Notice the strength of your alignment.
✔ *How does your shoulder feel?*
✔ *Are you holding it up?*
✔ *Are your arm and hand in alignment with your shoulder?*
✔ *Are they inside or outside the width of your shoulder?*
✔ *How does your elbow and wrist feel?*

figure 9.3

TIP Keep your hand and arm in alignment with your shoulder to lessen your muscular effort.

 Take a break before continuing with Part 2 on the next page.

Finding an effective angle of alignment: Part 2

 Stand again in a self-supported and comfortably aligned manner.

ACTION Once more apply pressure, with the top of your fist, to the side of your partner's back.

■ Feel the relationship of your body to your fist.
✔ *Does the angle or direction of your body support your fist?*

TIP Point your feet, pelvis and torso in the direction of your fist to help ensure that your body is not rotated away from your area of focus. (Just as you learned in lifting, facing your area of focus is important.)

ACTION For a moment, apply pressure using a non-supported angle of alignment.
✔ *How does this alignment compare to a supported one?* **(9.4)**

ACTION Practice applying pressure from different positions and finding the best angle of alignment. Also, explore using your right and left palm, forearm and elbow.

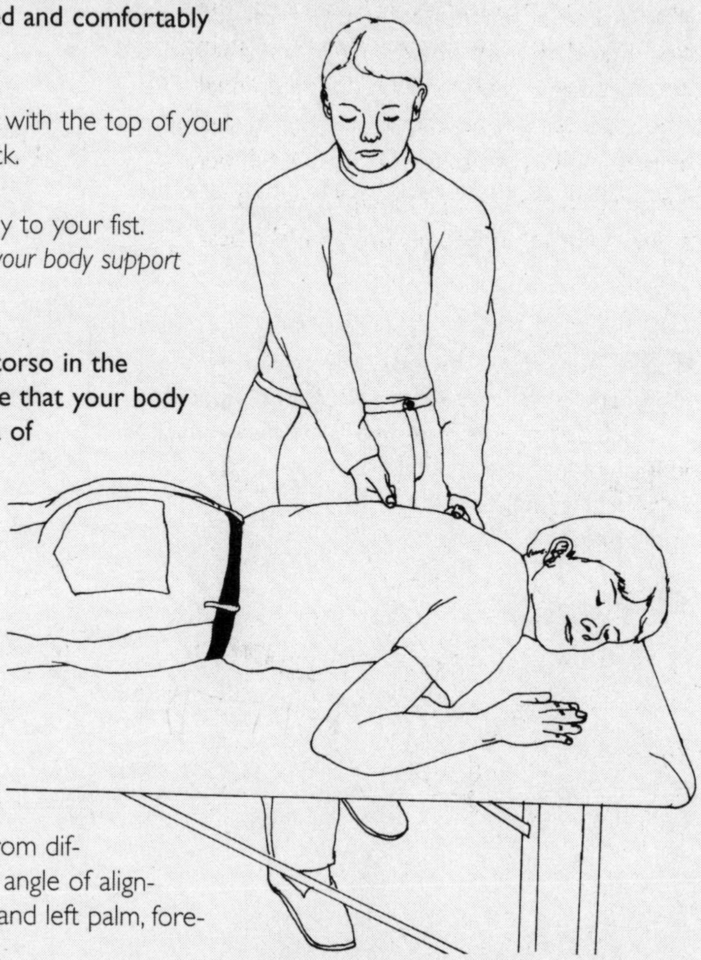

figure 9.4

 Ask your partner for feedback:
✔ *How did good alignment affect the application of pressure?*
✔ *How did poor alignment affect it?*
✔ *Which felt better to your partner? To you?*

Using Your Body Weight

It is very common for manual therapists to apply pressure by using sheer muscular strength. (**9.5a, 5b**) In fact, it is believed "the stronger the better" when it comes to applying pressure. Because of this belief, many manual therapists who are not "muscular" feel they can not adequately apply deep pressure. Well, here is some good news—if you weigh over 50 pounds, you can apply pressure with the best of them!

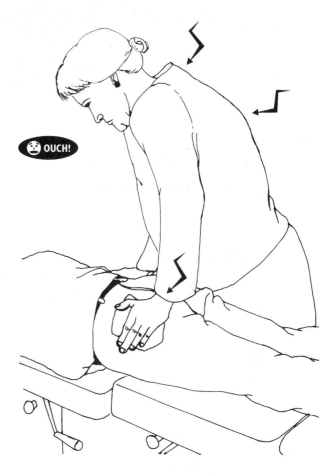

figure 9.5b

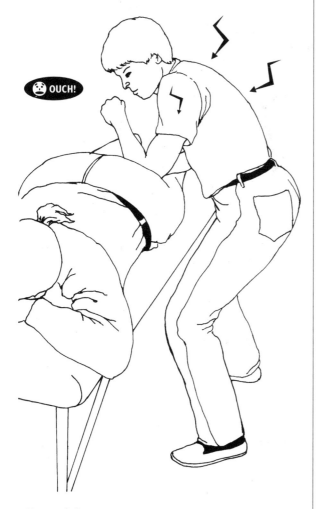

figure 9.5a

Think about how it would feel to have a 50 pound weight resting on your back. It would probably feel a little uncomfortable. The point is, in order to apply pressure effectively, you simply need to sink or drop your body weight into your area of focus and remain self-supported. (**9.6a, b**) Your muscles will naturally work to some degree, but you can effectively use your body weight to generate the depth you need to apply adequate pressure. Remember, you do not need to be a body builder to feel confident about applying pressure!

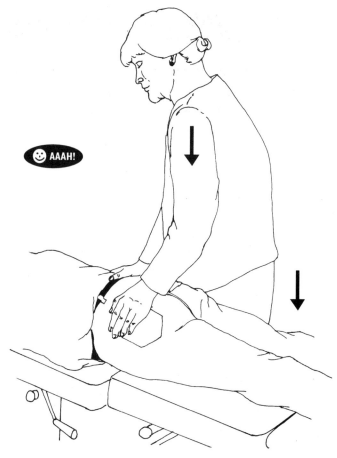

figure 9.6b

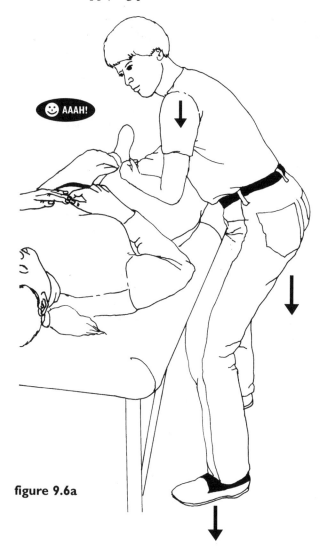

figure 9.6a

PREVENTION TIP

Breathing can help reduce discomfort and increase your overall energy. If you experience discomfort while working, consciously breathe into the area. This can help decrease your discomfort and increase your energy. Your breathing also reminds your client to breathe.

PARTNER PRACTICE

Using your body weight: Part 1

[11] With your partner lying prone or supine, stand close to your table, next to their leg.

ACTION Stand with your feet clearly connected to the ground, in a self-supported manner.

TIP Your feet's connection to the ground should be as clear as your elbow's connection to your partner.

ACTION Place your elbow gently on their buttocks, hamstrings, or quadriceps. **(9.7)**

ACTION Bend from your hip joints and slowly begin to bend your knees so that your weight drops or sinks into your elbow. This will increase the pressure you are applying.

TIP You can adjust the amount of pressure you apply by adjusting how much you bend from your hip joints and knees.

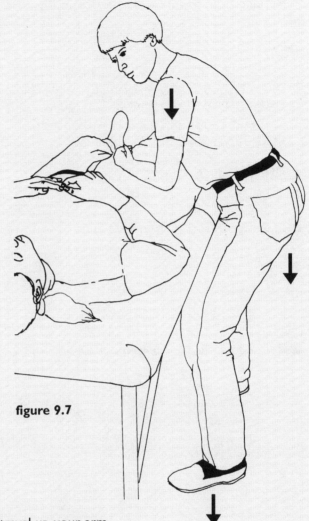

figure 9.7

ACTION As you increase the amount of pressure, keep your elbow and shoulder focused down into your pressure.

TIP Be careful not to let the pressure travel up your arm. This could cause your shoulder to rise up.

ACTION Hold the pressure for a moment and sense if you are using excessive muscular effort. If you are, see if you can reduce it by letting your body weight generate your strength.

ACTION Let your hand and arm relax and keep your back in a relaxed neutral position.

(STOP) *Take a break before continuing with Part 2 on the next page.*

Using your body weight: Part 2

Stand close to your table, next to your partner's leg.

ACTION Apply pressure again, but this time use excessive muscle strength and effort. **(9.8)**
- ✔ *How does this affect your shoulders and arms?*
- ✔ *Are you able to keep your back in a relaxed neutral position?*
- ✔ *What happens to your breathing?*

ACTION Use your body weight to apply the pressure.

■ Notice how using your body weight feels compared to using excessive muscular strength.

Ask your partner for feedback

- ✔ *How did using your body weight affect the amount of pressure you were able to apply?*
- ✔ *Did it feel as if you were able to control your pressure adequately?*
- ✔ *How did using your body weight compare to using your muscular strength?*

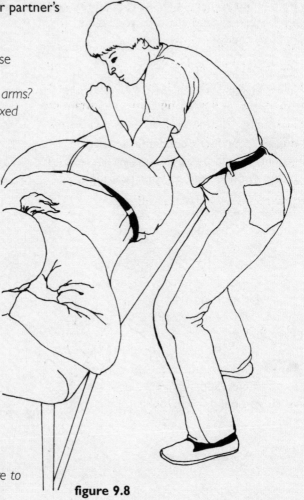

figure 9.8

Apply pressure by using your body weight, and you will find your muscles are less fatigued and sore, your joints less stiff and your energy level much higher throughout your day.

Get Closer

Whenever possible, position yourself close to your area of focus.

Stand or sit close to your table so you do not need to strain in order to reach your area of focus. Many times, for no reason other than habit, manual therapists find themselves standing or sitting several inches away from the area in which they are working. This puts strain on the entire body, especially the shoulders, arms and hands. (9.9) Working from a position where your upper body does not need to hold itself out in space decreases your fatigue and greatly reduces your overall effort.

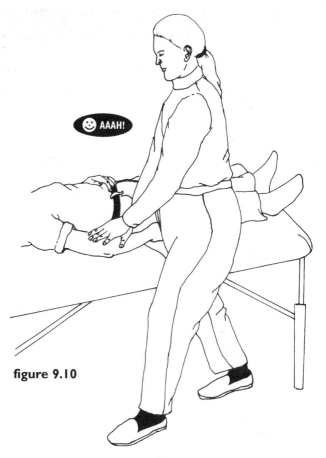

figure 9.10

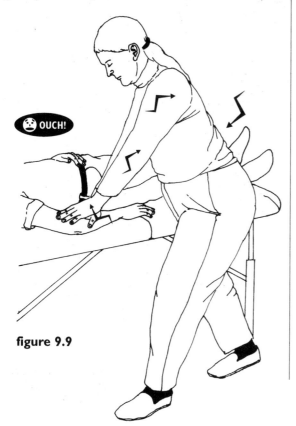

figure 9.9

Working close is especially important when it comes to applying pressure. Because applying pressure often involves holding the same position for several seconds, you need to keep your body as relaxed as possible. If you are standing several inches away and at the same time holding sustained pressure, you are asking your body to work overtime. There is no need to stand far away. Make it easy on yourself and stand close enough so you do not need to strain to reach your client. (9.10)

TRY THIS

Getting close

👣 Stand a few inches away from your table.

ACTION Apply some pressure to the edge of your table. **(9.11)**

TIP Stand far enough away so you need to reach out to apply the pressure. This might seem like an exaggeration, but it will make the point very clear.

✔ Can you feel how much effort it takes to reach your table from where you are standing?

✔ How stable and balanced are you?

✔ How do the muscles of your arms and shoulder feel?

✔ How about the muscles of your back and neck?

✔ From where do you bend in order to reach your table?

✔ Can you breathe easily from this position?

REST

👣 This time stand close to your table.

ACTION Apply pressure to the edge again. **(9.12)**

✔ Can you feel how standing close reduces the effort in your shoulders, arms and hands?

✔ How about the effort in your back and neck?

✔ Can you breathe more easily?

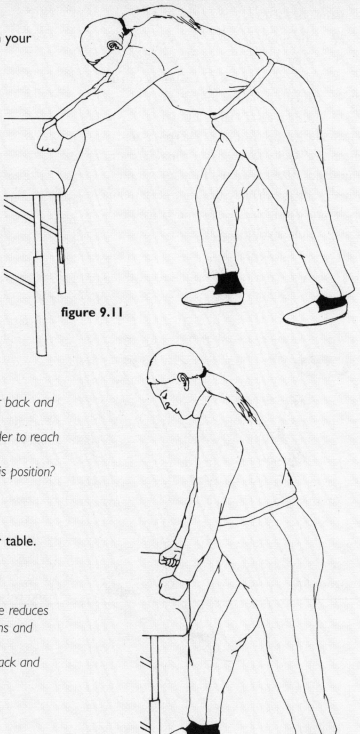

figure 9.11

figure 9.12

Believing in "Soft Tissue"

Often we alter our body mechanics based on beliefs we have about the needs of our client's body. For example, if your client tells you his back feels "like concrete," you might believe that it is going to take a lot of strength (and work!) to apply any kind of pressure to relieve this tightness. From this belief, then, you might do your work by contracting the muscles in your hands, forearms, and other areas of your body to apply the amount of pressure your think is necessary to achieve results.

Think about how you would apply pressure to a block of cement, or a big rock. You would probably stiffen your muscles and start to apply pressure with all of your physical strength and power. If you believe that your client's body is "hard as rock," "like concrete,"

or "tough as nails," you will use your body accordingly to apply pressure.

The human body is not like concrete—it is made up of soft tissue, minerals, salts and a whole lot of water. This makes us all very pliable and extremely sponge-like. If you keep this image in mind, it will help you to remember that you do not need to apply pressure like a jackhammer.

What we believe to be true about a person's body greatly influences how we approach them as manual therapists.

TRY THIS

A pillow of rock or feathers?

👣 Put a pillow on your table and stand next to it.

ACTION Look at this pillow for a minute and believe with all of your heart this pillow is as hard as a rock—this pillow is so hard you will never be able to apply enough pressure to make a dent into it. **(9.13)**

ACTION With this strong belief in your mind, begin to apply pressure to the pillow with your hands. **(9.14)**

■ Notice how you are using your hands and the quality of your muscle tone.
   ✔ *Are you using a lot of strength and power?*
   ✔ *Are the muscles in your hands and arms contracted?*

■ Notice how you are standing.
   ✔ *Are you standing in a self-supported manner?*
   ✔ *Are you breathing normally?*

REST

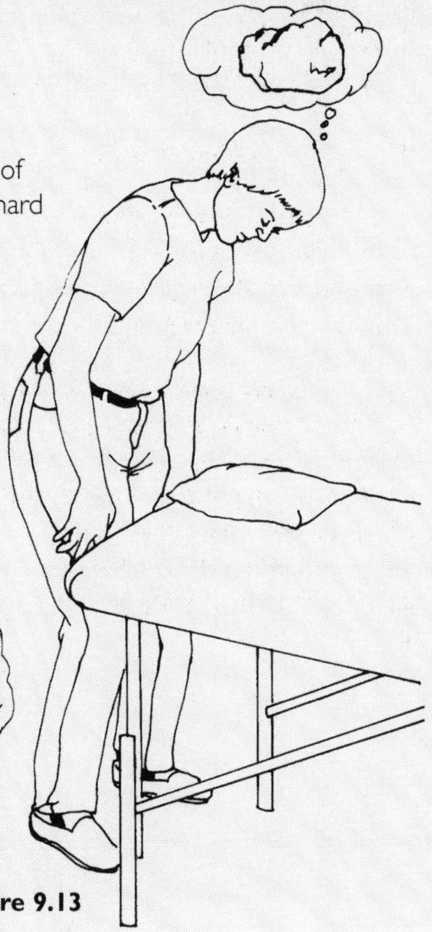

figure 9.13

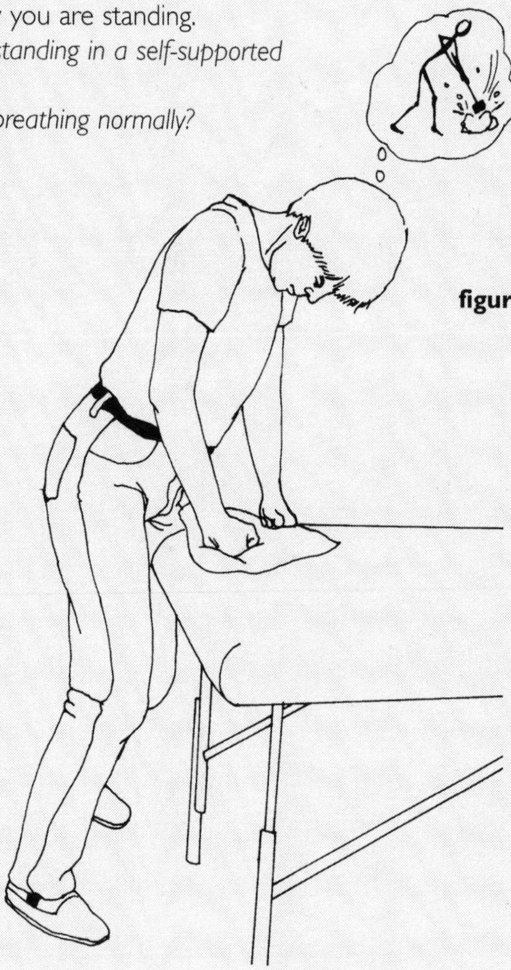

figure 9.14

Stand and look at the pillow again, but this time see the pillow for what it is, soft and pliable. **(9.15)**

ACTION With this belief in your mind, begin to apply pressure to the pillow with your hands. **(9.16)**

■ Notice if you are using the same amount of strength as before.
   ✔ *How do the muscles in your hands and arms feel?*
   ✔ *Are they contracted or are they a little more relaxed?*
   ✔ *Are you standing in a more self-supported manner?*
   ✔ *Are you breathing normally?*

When applying pressure to a soft and pliable pillow, your hands can remain soft, your muscular effort can be less and your breathing can remain normal. The same is true when applying pressure to your client. Become aware of how your beliefs influence your body mechanics, and how believing that the body is soft and pliable can decrease your strain and effort!

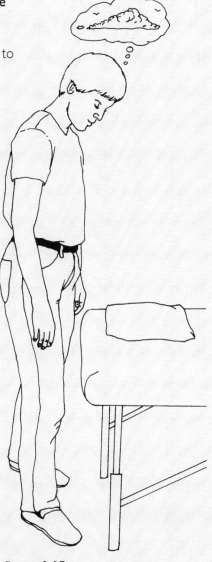

figure 9.15

figure 9.16

Rigor Mortis Anyone?

Holding or applying pressure in one area for several seconds or minutes is common in manual therapy. This can often be the cause of stress and extreme fatigue for the practitioner. Just because you are applying "static" pressure, it does not mean your entire body must become "static" as well. It is all too common for manual therapists to begin applying static pressure and slowly begin to look as if rigor mortis is settling in. (9.17)

As we have said, reducing your muscular effort will decrease stiffness and fatigue. The less isometric contractions your muscles hold, the more energy and vitality they will have. This is why using your body weight is so important. The other important point to remember is—keep breathing!

Often therapists feel faint and quickly lose energy while applying sustained pressure. This is because their breathing is restricted. When the muscles are being held in isometric contraction, the normal response is to alter the breathing and contract the diaphragm. Breathe consciously when you are applying sustained pressure and keep the rest of your body alive and moving.

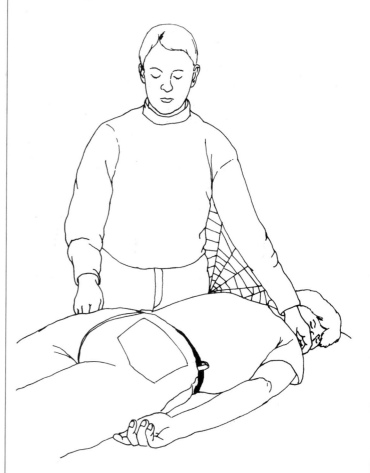

figure 9.17

PARTNER PRACTICE

Breathing while applying static pressure: Part 1

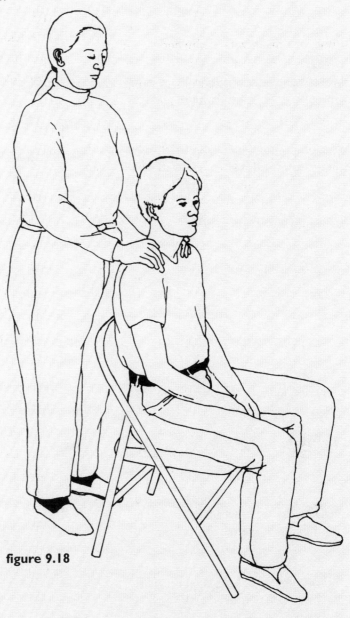

Have your partner sit in a chair. Stand close behind them.

ACTION With your hands, apply pressure to your partner's shoulder and hold it for about 30 seconds. **(9.18)**

- Notice how much effort you are using.
    - ✔ Are you tightly holding the muscles of your hands, arms and shoulders?
    - ✔ Are you using your body weight to apply the pressure?

- Notice how you are breathing.
    - ✔ Are you breathing deeply?
    - ✔ Shallowly?

REST

figure 9.18

198

Breathing while applying static pressure: Part 2

ACTION Standing behind your partner, apply and hold pressure to their shoulders once again.

TIP Focus on using your body weight and less muscular effort in the muscles of your hands, arms and shoulders.

■ As you hold the pressure, begin to consciously breathe slowly and take a few deep breaths. **(9.19)**
  ✔ *Can you feel a part of yourself moving as you breathe?*
  ✔ *What part do you feel moving?*
  ✔ *Your chest?*
  ✔ *Abdomen?*

ACTION Continue to breathe slowly and allow the movement of your breath to move your shoulders and arms.

TIP Let the joints of your shoulders and elbows respond to the movement of your breath.

ACTION As you breathe, let your hip joints and knees respond to the movement and rhythm of your breathing.

TIP You may feel as if your knees and hip joints are moving up and down.

ACTION Let your head, jaw, neck and back also follow the movement of your breathing.

■ While consciously breathing, sense how your entire body can be very alive and dynamic as you apply and hold pressure.

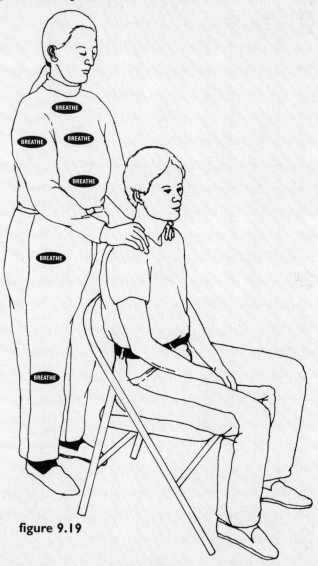

figure 9.19

ACTION Repeat this lesson using your elbow or forearm.

Ask your partner for feedback.

✔ *How does your breathing affect the application of pressure?*
✔ *Did your breathing encourage them to breathe as well?*

APPLYING PRESSURE SUMMARY

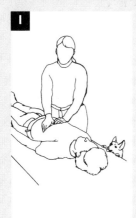

1

Find an effective angle of alignment.
An effective angle of alignment is one in which you can transfer your body weight to your shoulders, to your elbows, to your wrist, ultimately feeling the support of your joints.

2

Use your body weight.
You can apply pressure effectively by bending from your hip joints and knees and sinking your body weight into your area of focus.

3

Get close.
Whenever you have the opportunity, position yourself close to your area of focus. This means standing or sitting close to your table so you do not need to strain in order to reach your area of focus.

4

Believe in "soft tissue."
The human body is not like concrete. What we believe to be true about a person's body greatly influences how we approach them as manual therapists.

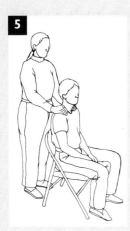

5

Breathe while applying "static" pressure.
Breathe consciously when applying sustained pressure to keep the rest of your body alive and moving.

Notes:

Notes:

Notes:

About Those Animals...

Lop-eared Rabbit

Domestic Shorthair Cat

Pembroke Corgie

Rufus Hummingbird

Impala

Collared Lemming

Gibbon

Chinook Salmon

Shark

Blackfooted Ferret

Wasabi

Bibliography

[1]Rosenfeld, Edward. The Forebrain: Sleep, Consciousness, Awareness and Learning: An Interview with Moshe Feldenkrais, *Interface Journal*, Nos. 3-4. 1976.

Bailey, Donna. *Track and Field.* Austin: Steck-Vaughn Co., 1991.

Bean, Constance A. *The Better Back Book.* New York: William Morrow and Company, Inc., 1989.

Biel, Andrew. *Trail Guide to the Body: How to Locate Muscles, Bones and More!* Boulder: Andrew Biel, 1997.

Brennan, Richard. *The Alexander Technique Workbook.* Shaftesbury: Element Books Limited, 1992.

Cailliet, Rene. *Neck and Arm Pain.* Philadelphia: F. A. Davis Company, 1981.

Cech, Donna. *Functional Movement Development Across the Life Span.* Philadelphia: W.B. Saunders Co., 1995.

Dunlap, Knight. *Habits: Their Making and Unmaking.* New York: Liveright, 1972.

Erickson, Milton H. *Life Reframing in Hypnosis.* New York: Irvington Publishers, Inc., 1985.

Falk, Dean. *Braindance.* New York: Henry Holt and Co. Inc., 1992.

Fash, Bernice. *Body Mechanics in Nursing Arts.* New York: McGraw-Hill Book Co., 1946.

Feldenkrais, Moshe. *Adventures in the Jungle of the Brain: The Case of Nora: Body Awareness as Healing Therapy.* New York: Harper & Row, 1977.

Feldenkrais, Moshe. *Awareness Through Movement: Health Exercises for Personal Growth.* New York: Harper & Row, 1972.

Feldenkrais, Moshe. *Body and Mature Behavior: A Study of Anxiety, Sex, Gravitation & Learning.* Madison: International Universities Press, 1949.

Florentino, Mary R. *A Basis for Sensorimotor Development–Normal and Abnormal.* Springfield: Charles C. Thomas, 1981.

Frost, Loraine. *Posture and Body Mechanics.* Iowa City: State University of Iowa, 1952.

Gaskin, John. *Movement.* New York: F. Watts, 1984.

Germain-Calais, Blandine, Lamotte, Andree. *Anatomy of Movement: Exercises.* Seattle: Eastland Press, 1996.

Gibson, Gary. *Pushing and Pulling.* Connecticut: Copper Beech Books, 1995.

Goldthwait, Joel E. *Essentials of Body Mechanics in Health and Disease.* Philadelphia: J. B. Lippincott, 1952.

Greene, Lauriann. *Save Your Hands! Injury Prevention for Massage Therapists.* Seattle: Infinity Press, 1995.

Grieve, June I. *Muscles, Nerves and Movement: Kinesiology in Daily Living*, 2nd edition. London: Blackwell Science Ltd., 1996.

Hall, Carrie M., Brody, Lori T. *Therapeutic Exercises; Moving Toward Function.* Philadelphia: Lippincott Williams & Wilkins, 1999.

Hall, Mina. *The Big Book of Sumo.* Berkeley: Stone Bridge Press, 1998.

Hall, Susan J. *Basic Biomechanics.* New York: McGraw-Hill Co., 1998.

Haller, Jeff. *Use of Self* (audio tapes) Trelleborg, Sweden: Haller, 1997.

Hoppenfield, Stanley. *Physical Examination of the Spine and Extremities.* Norwalk: Appleton & Lange, 1976.

Jacobs, Karen. *Ergonomics for Therapists.* Oxford: Butterworth-Heinemann, 1999.

Jenkins, David B. *Hollingshead's Functional Anatomy of the Limbs and Back.* Philadelphia: W. B. Saunders Co., 1991.

Juhan, Dean. *Job's Body: A Handbook for Bodywork.* Barrytown: Station Hill, 1987.

Kapandji, I. A. *The Physiology of the Joints.* New York: Church Livingstone Inc., 1987.

Kelly, Ellen D. *Teaching Posture and Body Mechanics.* New York: A.S. Barnes, 1949.

Kendall, FP, McCreary, EK. *Muscles: Testing and Function*, 3rd edition. Baltimore: Williams & Wilkins, 1983.

Lederman, Eyal. *Fundamentals of Manual Therapy: Physiology, Neurology and Psychology.* New York: Churchill Livingstone, Inc., 1997.

Lee, Mabel. *Fundamentals of Body Mechanics & Conditioning: An Illustrated Teaching Manual.* Philadelphia: W. B. Saunders, 1949.

Linden, Paul. *Compute in Comfort.* Upper Saddle River: Prentice Hall PTR, 1995.

Lippert, Herbert. *Anatomie: Text und Atlas.* Berlin: Urban & Schwarzenberg, 1976.

Malalasekera, G. P. *Encyclopedia of Buddhism.* Colombo: Government of Ceylon, 1973.

Martin, Suzanne. *Functional Movement Development Across the Life Span.* Philadelphia: W.B. Saunders Co., 1995.

Mayglothling, Rosie. *Rowing.* Wiltshire: The Crowood Press, 1990.

McMinn, R.M.H. *Color Atlas of Human Anatomy.* Chicago: Year Book Medical Publishers, 1985.

Mitchell, Stewart. *The Complete Illustrated Guide to Massage.* Great Britain: Element Books Limited, 1997.

Mumford, Susan. *Healing Massage: A Practical Guide to Relaxation and Well-Being.* New York: Penguin Putman Inc., 1997.

Peterson, Roger T. *Field Guide to Animal Tracks (Peterson Field Guides).* Oxfordshire: Houghton Mifflin Co., 1998.

Platzer, Werner. *Color Atlas and Textbook of Human Anatomy*, Volume 1: Locomotor System, 3rd edition. New York: Thieme Inc., 1986.

Reynolds, Edward. *The Evolution of the Human Pelvis in Relation to the Mechanics of the Erect Posture.* Cambridge: The Museum, 1931.

Roaf, Robert. *Posture.* New York: Academic Press, 1977.

Roberts, Tristan, D. M. *Understanding Balance: The Mechanics of Posture and Locomotion.* New York: Chapman & Hall, 1995.

Rolf, Ida P. *Rolfing: Reestablishing the Natural Alignment and Structural Integration of the Human Body.* Rochester: Healing Arts Press, 1989.

Rolf, Ida. *Rolfing and Physical Reality.* Rochester: Healing Arts Press, 1990.

Rolf, Ida. *Rolfing: Integration of Human Structures.* New York: Harper Row, 1977.

Salvo, Susan G. *Massage Therapy: Principles and Practice.* Philadelphia: W.B. Saunders, 1999.

Schumacher, John A. *Human Posture: The Nature of Inquiry.* New York: State University of New York Press, 1989.

Sechi, Davide. *Massage Basics.* New York: Sterling Publishing Co., 1998.

Shafarman, Steven. *Awareness Heals: The Feldenkrais Method for Dynamic Health.* Massachusetts: Addison-Wesley Publishing Co., 1997.

Sivananda Yoga Vedanta Center. *Yoga Mind and Body.* New York: Dorling Kindersley, 1996.

Staub, Frank. *Mountain Goats.* Minneapolis: Lerner Publications Co., 1994.

Tappan, Frances M., Benjamin, Patricia J. *Tappan's Handbook of Healing Massage Techniques*, 3rd edition. Stamford: Appleton & Lange, 1998.

Thompson, Clem W. *Manual of Structural Kinesiology*, 11th edition. St. Louis: Times Mirror/ Mosby College, 1989.

Thompson, Diana L. *Hands Heal: Documentation for Massage Therapy.* Seattle: Diana Thompson, 1993.

Todd, Mabel E. *The Thinking Body.* Brooklyn: Dance Horizons, 1979.

Tortora, Gerald. *Principles of Human Anatomy*, 4th edition. New York: Harper & Row, 1986.

Tozeren, Aydin. *Human Body Dynamics: Classical Mechanics and Human Movement.* New York: Springer-Verlag, 1999.

Tyldesley, Barbara. *Muscles, Nerves and Movement: Kinesiology in Daily Living*, 2nd edition. London: Blackwell Science Ltd., 1996.

Vogel, Steven. *Cats' Paws and Catapults: Mechanical Worlds of Nature and People.* New York: W. W. Norton & Co., 1998.

Winters, Margaret C. *Protective Body Mechanics in Daily Life and in Nursing.* Philadelphia: W. B. Saunders, 1952.

Zacharkow, Dennis. *Posture: Sitting, Standing, Chair Design and Exercise.* Springfield: Charles C. Thomas, 1988.

Zacharkow, Dennis. *The Healthy Low Back.* Springfield: Charles C. Thomas, 1984.

World Wide Web Sources

Back and Neck Care Guide. McKinley Health Center, University of Illinois. February 8, 2000 Retrieved: March 3, 2000 from the World Wide Web: http://www.uiuc.edu.departments/ mckinley/health-info/fitness/back

Die Entwicklung im 1. Lebensjahr Retrieved: April 3, 2000 from the World Wide Web: http://www.home.wtal.de/uerhage/ babyentwicklung.htm

Southeastern Hand Center. Surgery of the Hand and Upper Extremity. Jack L. Greider. Retrieved: April 17, 2000 from the World Wide Web: http://www.handsurgery.com

Study: Human Ancestors had Knuckle-Walking Characteristics. March 22, 2000. Associated Press. Retrieved: April 17, 2000 from the World Wide Web: http://www.cnn.com/2000/nature/03/22/ knuckle.walkers.ap

Index

Notes:

Notes:

Order Form

Call toll free and order now!

Telephone orders:

Toll Free: (877) 474-1001
Please have your VISA or MasterCard ready.

Postal orders:

Mail to: FRYETÄG Publishing, P.O. Box 2688, Stanwood, WA, 98292

Please send_____copy(s) of *Body Mechanics for Manual Therapists* ($38.95 each)

Company Name: _____

Name:_____

Address: _____

City: _____ State: _____ Zip:_____

Telephone: (____) _____

Sales Tax:

Please add 8.6% for books shipped to WA addresses.

Shipping:

$5.00 for the first book and $1.50 for each additional book. Shipping may take two to four weeks. Bulk order discounts available.

Sub total _____

Payment:

❏ Check (payable to FRYETÄG Publishing)
❏ Credit card: ❏ VISA ❏ MasterCard

Tax _____

Shipping_____

TOTAL_____

Card number: _____

Name on card: _____ Exp. date: _____